IMAGES
of America

GRISTMILLS OF CENTRAL TEXAS

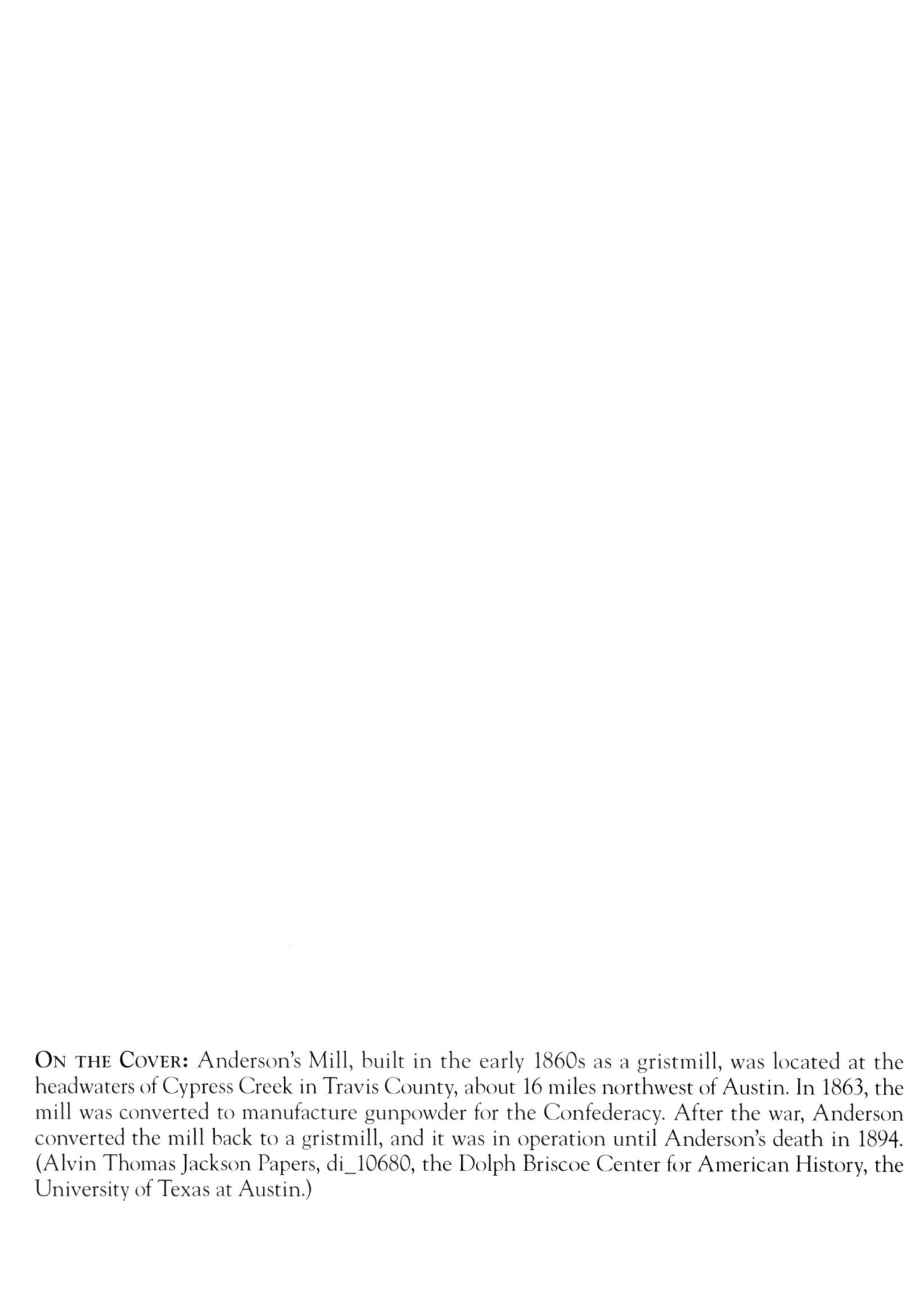

ON THE COVER: Anderson's Mill, built in the early 1860s as a gristmill, was located at the headwaters of Cypress Creek in Travis County, about 16 miles northwest of Austin. In 1863, the mill was converted to manufacture gunpowder for the Confederacy. After the war, Anderson converted the mill back to a gristmill, and it was in operation until Anderson's death in 1894. (Alvin Thomas Jackson Papers, di_10680, the Dolph Briscoe Center for American History, the University of Texas at Austin.)

IMAGES
of America

GRISTMILLS OF CENTRAL TEXAS

Charlene Ochsner Carson

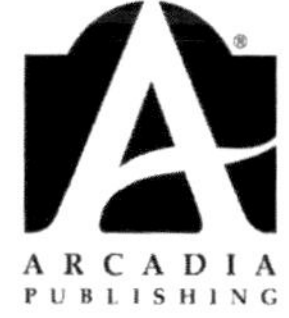

ISBN 978-1-4671-2596-3

Published by Arcadia Publishing
Charleston, South Carolina

Printed in the United States of America

Library of Congress Control Number: 2016958661

For all general information, please contact Arcadia Publishing:
Telephone 843-853-2070
Fax 843-853-0044
E-mail sales@arcadiapublishing.com
For customer service and orders:
Toll-Free 1-888-313-2665

Visit us on the Internet at www.arcadiapublishing.com

To the millwrights and millers of Central Texas—the true heroes of the untamed Texas frontier

Contents

Acknowledgments

A special thank you to everyone who made this book possible: my husband, Maurice, who traveled the backroads of Texas with me searching for mills and who was my photographer and number one supporter; my son, David Carson, for his computer and editorial skills; my daughter, Donna Johnson; and grandson, Dashiell Johnson, for their encouragement and support; and dear friends Nancy Shepperd for photograph restoration and Carol Wilson for proofreading the manuscript.

Thanks to Brandon Aniol for providing a tour of the Landmark Inn site; Shahar Yarden for providing a tour of the Homestead Heritage site; Carl Chisum for providing a tour of the Rio Frio Mill site; Mac Hickerson for providing a tour of the Mote Smith Mill site; Bob Dube for providing a tour of the Chadwick Mill site; Lawrence Dube for sharing information regarding the Chadwick Mill; Olaf Harris, owner of Norway Mills in Clifton, for allowing us to photograph the mill on his property; Ed Bell for escorting us to the Norway Mills site; the Hill family for allowing us to tour their Stinnett Mill home; the Paul Myers Foundation for allowing us to tour Summers Mill; Dayton Regian for escorting us on the Summers Mill site; Ashley Unger for technical assistance; Michael and Nancy Kelsey for tips on where to find information and for photographs; and the many others who contributed photographs—their names appear with their photographs.

I am also grateful to the hundreds of librarians and archivists throughout Central Texas who responded to my requests for information and photographs. Everyone responded in a timely manner and with generous hearts.

A special acknowledgment goes to A.T. Jackson for his book *Mills of Yesteryear* (Texas Western Press, the University of Texas at El Paso, 1971). Mr. Jackson's book was my primary source of information on the history of milling in Central Texas.

Introduction

The first gristmills in what is now known as Texas were built when the king of Spain authorized the founding of missions in the area under control of the Spanish government. One of those was the San Jose Mission, located at San Antonio near the San Antonio River. This mission, founded in 1720, is the largest of the Texas missions. A gristmill was added to the complex around 1794, when the Spanish government issued orders that wheat, in addition to corn, be cultivated at the missions. Another gristmill authorized by the king of Spain was the Molino Blanco, meaning "white mill." This mill was built near the Alamo in 1733 and operated until about 1830, providing almost a century of service.

In 1821, Mexico, having been under Spanish rule since 1535, won independence from Spain. Texas then became a part of the new Empire of Mexico, with the Mexican flag flying over it. In 1820, Moses Austin asked Spanish officials to let him establish a colony of Americans in Texas. His request was granted, but Austin died before he could organize the colony. In 1823, the Mexican Congress confirmed Austin's land grant, and his son, Stephen F. Austin, carried out his father's plans and brought 300 families into Texas.

During that time, other *empresarios* had received land grants from Mexico to establish colonies. The Mexican government thought it was in its best interest to populate the regions between Mexico and the Indians to the north and the expanding United States on the east. During the next few years, the empresarios founded many colonies and brought several thousand settlers into Texas. Land grants to encourage the building of mills were not uncommon.

Mexican government officials became alarmed at the increasing number of emigrants from the United States, so in 1830, they halted American immigration to Texas. Four years later, in 1834, Gen. Antonio Lopez de Santa Anna, a Mexican soldier and politician, overthrew Mexico's constitutional government and made himself dictator, creating an unstable political situation throughout Mexico, including Texas. Texans felt that their personal and economic security was threatened, and they became alarmed about their future well-being. Therefore, in 1835, the colonists in Texas revolted against Mexico, and the Texas Revolution began with the Battle of Gonzales on October 2, 1835.

The Texans fought in several clashes and battles; the most noted was the Battle of the Alamo. Santa Anna began his 13-day siege of the Alamo on February 23, 1836. While the Alamo was surrounded by Santa Anna's troops, Texas government officials, meeting at Washington-on-the Brazos on March 2, 1836, officially declared independence from Mexico. The Alamo fell to Santa Anna's troops four days later, March 6, 1836.

The devastating defeat at the Alamo was a wake-up call to Texans. They had land, homes, families, and political and economic freedom to defend. They eagerly joined Sam Houston's forces as he marched toward San Jacinto in pursuit of Santa Anna. The revolution came to an end when Sam Houston's army defeated the Mexicans in the Battle of San Jacinto on April 21, 1836. Soon thereafter, Texas became the independent Republic of Texas.

It would be another 10 years, however, before the question of who owned Texas was settled. The entry of Texas into the Union in December 1845 touched off the Mexican-American War. Even though Texas had defeated the Mexican troops at San Jacinto, Mexico refused to recognize the independence of Texas. After several Mexican invasions of Texas, Pres. James K. Polk signed a resolution declaring a state of war between the United States and Mexico.

The war ended in March 1847 with the capture of Mexico City. According to the terms of the Treaty of Guadalupe-Hidalgo, signed on February 2, 1848, Mexico, for the first time, renounced all claims to Texas and accepted the Rio Grande as the boundary between the United States and Mexico. The question of who owned Texas had been settled. Emigrants from European countries as well as people from other states began flooding into the young state.

Land where fresh water springs gushed and flowed was in high demand, despite the fact that the area was occupied by hostile Indians who had been there for centuries and unfriendly bears, wildcats, and other animals. Among the bravest of the brave who ventured into this unknown frontier were the millwrights who were attracted by the available waterpower of the untamed rivers and streams that flowed throughout the landscape of Texas. Some had worked as millers in Europe and had come to North America seeking political freedom and economic opportunity. Several had served the Republic of Texas in its war for independence with Mexico; some received land grants for their service to the republic. Others served with the militia, protecting the influx of white settlers coming into Texas from Indian attacks. Some were slaveholders who brought their slaves with them and then used the slave labor to build their mills.

Most millers enjoyed a reputation of being among the most respected people in the community. They were honest and trustworthy in their business dealings. Many served their community as school trustees, civic leaders, and community builders. Most were innovative and had natural engineering and mechanical skills. Supplies that were not readily available were made by hand out of raw materials that were obtainable from the land.

Most mills in Texas were built between 1840 and the late 1870s. Those that were in operation during the Civil War played a key role in supporting the Confederate army. Some mills were commissioned to card wool for the making of Confederate uniforms. Others were commandeered to make ammunition, and still others provided flour for the troops. Millers were exempt from military service so that they could focus on operating the mills. Some mills were built like forts, complete with gun ports, as a safe haven for white settlers. The primary and most important purpose of the gristmill, however, was to grind wheat into flour and corn into meal for the settlers' daily bread.

One

The San Antonio River
Home of the Spanish Missions

The San Antonio River rises in a cluster of springs about four miles above the city. One group is known as the San Antonio Springs, and another nearby is known as the San Pedro Springs. When the springs unite, they form a bold, rapid stream through the heart of the city. Eventually, the stream becomes a river that follows a southeastern path through the state and ultimately feeds into the Guadalupe River about 10 miles northwest of the San Antonio Bay. Its principal tributaries are the Medina River, which flows through Castroville, site of the Landmark Inn and mill, and Cibolo Creek, which flows through Boerne, site of the William Dietert Mill.

Native American hunter-gatherers were the first people to utilize the springs and the river they created. The next group to lay claim to the area was Spanish explorers, followed by Franciscan missionaries.

When Franciscan Antonio Olivares arrived with a group of Spanish explorers, he immediately recognized the beauty of the river and began a nine-year campaign to build a mission on its bank. In 1718, Olivares built a mission near the headwaters of the San Antonio River. He called it San Antonio de Valero. Relocated in 1719 and again in 1739, the mission finally came to rest on the east bank of the San Antonio River. Today this mission is known as the Alamo.

During the remainder of the 18th century, four other major Spanish missions occupied sites along the historical course of the San Antonio River, including Mission San Jose, established in 1720. The other three missions, Mission Concepcion, Mission San Juan Capistrano, and Mission Espada, were established in East Texas and were moved to San Antonio in the early 1700s.

A gristmill known as Molino Blanco was built on the San Antonio River near the Alamo in 1733 and operated until about 1830. Other mills associated with the San Antonio River, included in this chapter, are Berg's, Nathaniel Lewis, Jacob Laux, and C.H. Guenther. The Frio River, west of San Antonio, was home to the Rio Frio Mill and Gin at Leakey, and the Rio Grande was home to the Simeon Hart Mill.

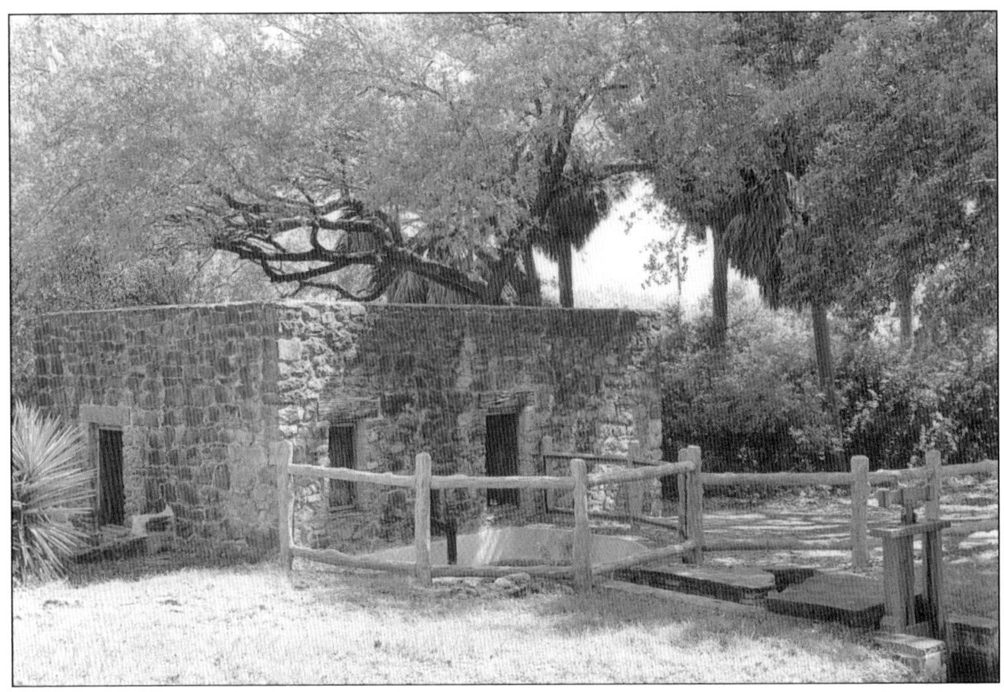

Mission San Jose y San Miguel de Aguayo in San Antonio, Bexar County, was founded in February 1720. The building of the baroque limestone church began in 1768. At that time, 350 Indians were residing at the mission. Franciscans and the Indian inhabitants built the Spanish colonial gristmill, located north of the church building, when orders were given to cultivate wheat in addition to corn around 1794. Hand-dug irrigation canals called *acequias* brought water from the San Antonio River to the mill. The mill operated until the early 1800s. Over time, the mill fell into a state of disrepair, and it was forgotten about until the 1930s, when workers from the Works Projects Administration accidentally stumbled upon the wheel room and forebay when restoring the mission complex. The mill was restored to working order and rededicated in 2001 as the oldest operating mill in Texas. (Author's collection.)

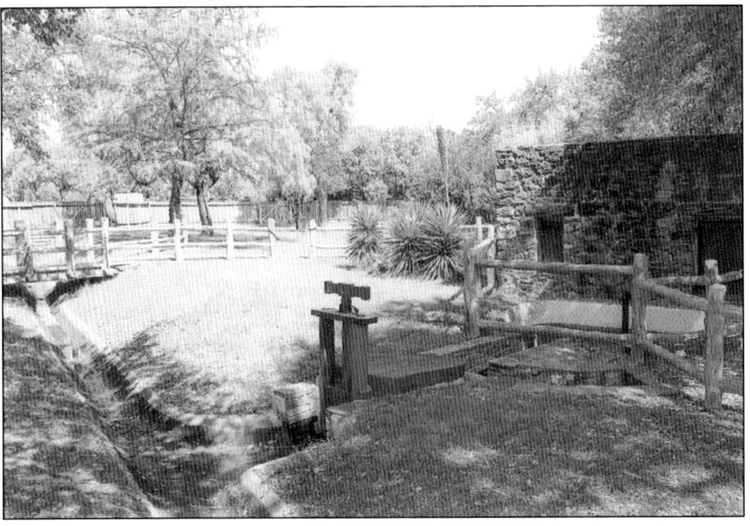

Water for the mill flows from a section of the Mission San Jose acequia through an authentically reproduced sluice gate into a 12-foot-deep conical pit, or equalizer vat, until it is released to power the mill. The vertical sluice gate is used to regulate the flow of water into the pit. (Author's collection.)

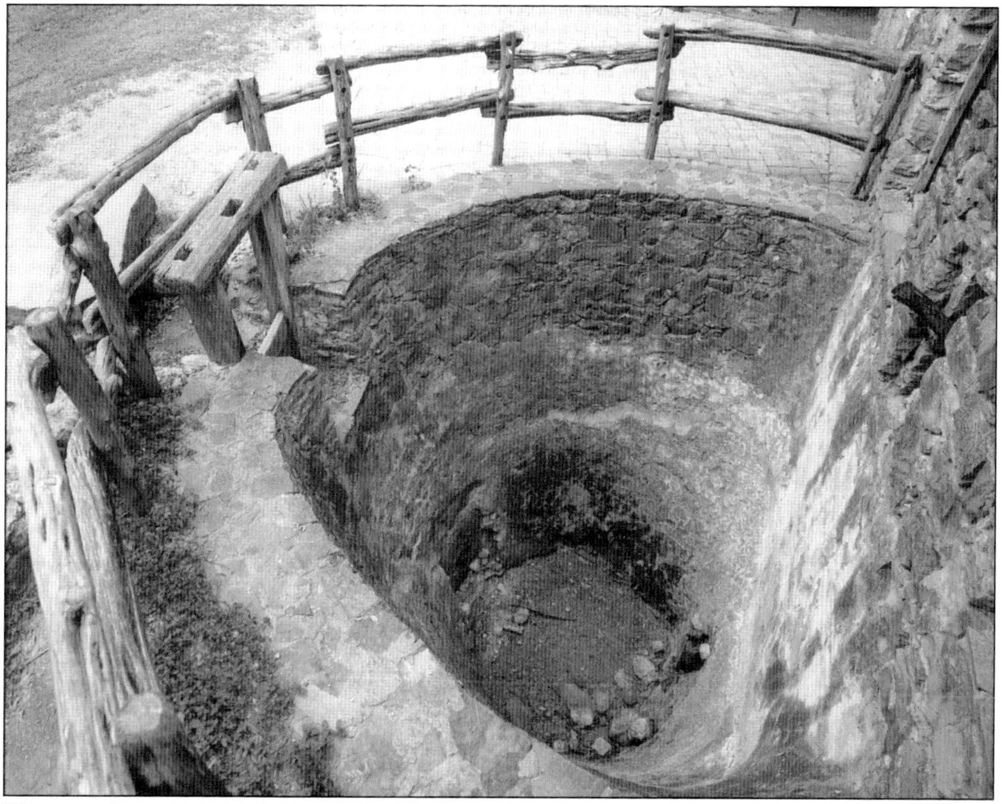

This is a close-up view of the conical pit, or equalizer vat, where water is held prior to powering the flutter wheel. From the bottom of this pit, a flume directs the water onto the turbine blades, striking them with sufficient force to turn a horizontal wheel and shaft connected to grinding stones in a room above the wheel. (Library of Congress, Prints & Photographs Division, Historic American Engineering Record [HAER] TEX, 15-SANT,30-5.)

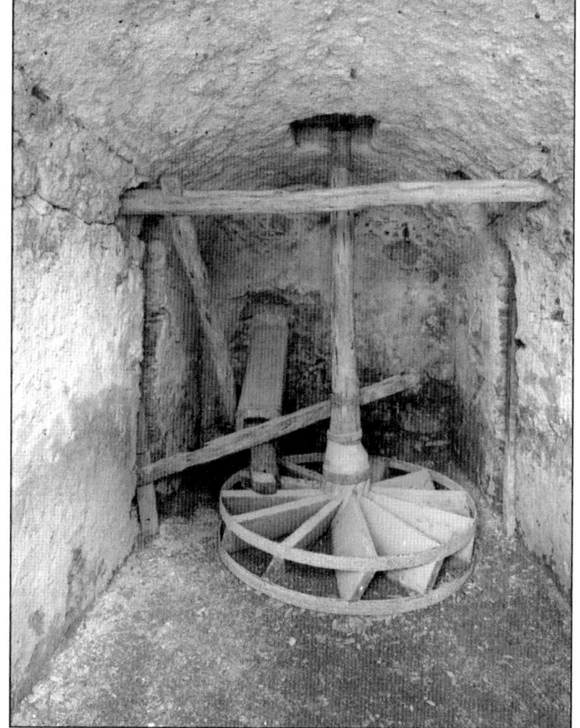

This is a view of the flutter wheel in the basement of the San Jose gristmill. The rectangular enclosed flume protruding from the wall at center left carries the water from the equalizer vat that powers the wheel. As the wheel rotates, the shaft turns the millstone on the floor above. (Library of Congress, Prints & Photographs Division, HAER TEX, 15-SANT,30-6.)

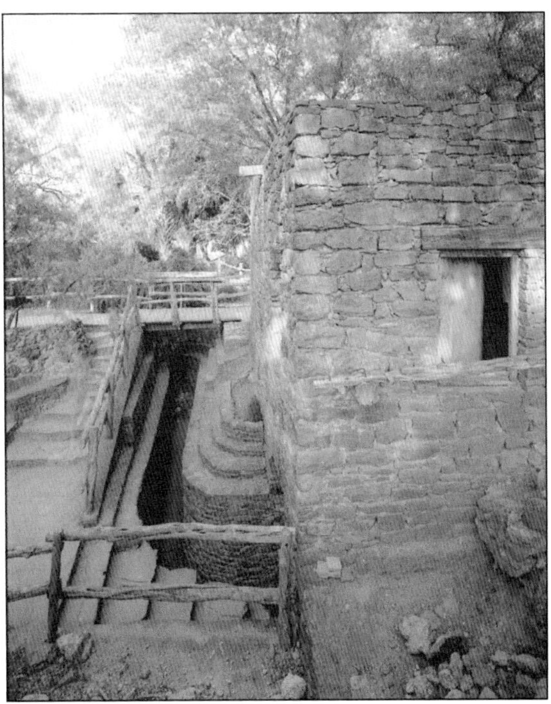

Water is then discharged down the tailrace, shown at left, on the backside of the mill building. The water follows a channel meeting the acequia at a lower elevation downstream. The discharged water would traditionally have been used to irrigate the fields outside the mission. (Library of Congress, Prints & Photographs Division, HAER TEX, 15-SANT,30-4.)

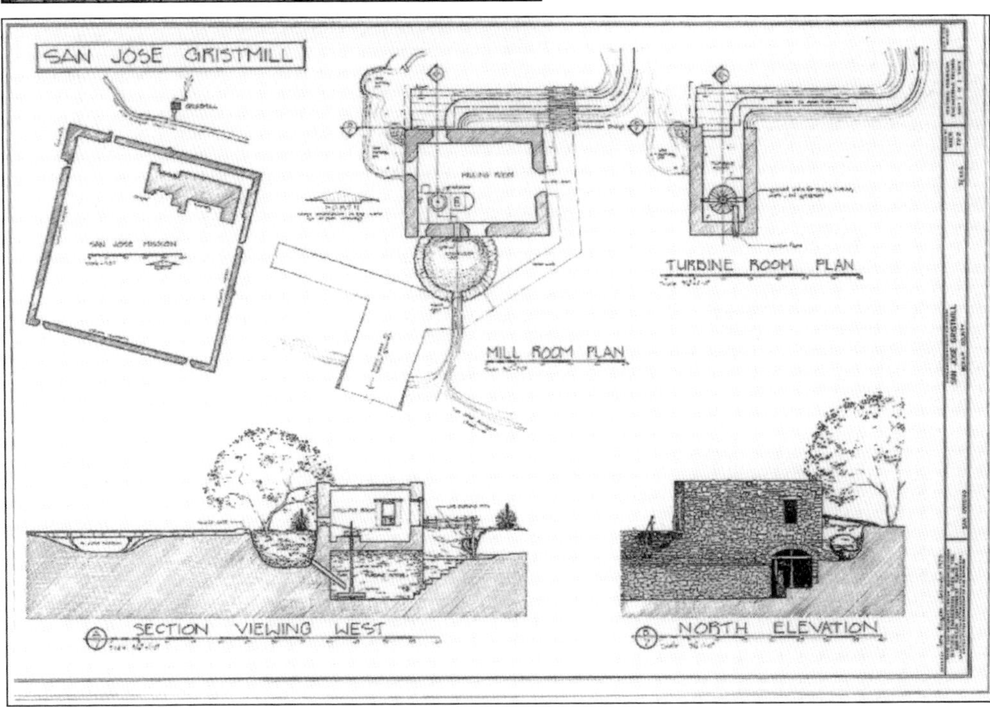

These engineering drawings show the overall water flow process from an overhead view. Water enters the inlet flume, filling the tub, then on to the waterwheel, and finally empties through the tailrace. The turbine room is located directly below the mill room. The section view (bottom left) looks at the same process from the side and provides a clear picture of the water elevation changes. (Library of Congress, Prints & Photographs Division, HAER TEX, 15-SANT,30.)

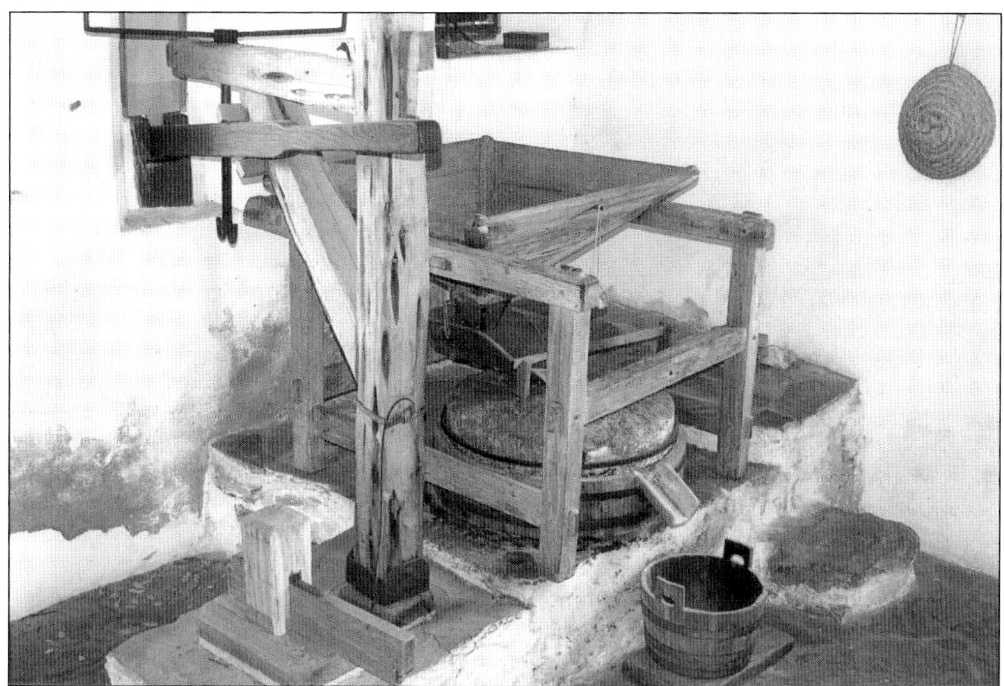

These photographs offer two different views of the interior of the San Jose millhouse. To operate the mill, grain is poured into the hopper, where it falls through the center hole, or eye, of the millstone. The top stone, called the runner, turns and grinds the grain against the stationary bottom stone, called the bedstone. Grooves, or furrows, carved in both stones push the grain to the outside edge of the stones. The furrows on the top stone cross the furrows on the bottom stone like a pair of scissors. By the time the grain has worked its way out from the center, it has been ground into meal, and it is channeled through the spout into a container. When the mill is working at full capacity, it can grind one bushel of grain every hour. (Both, author's collection.)

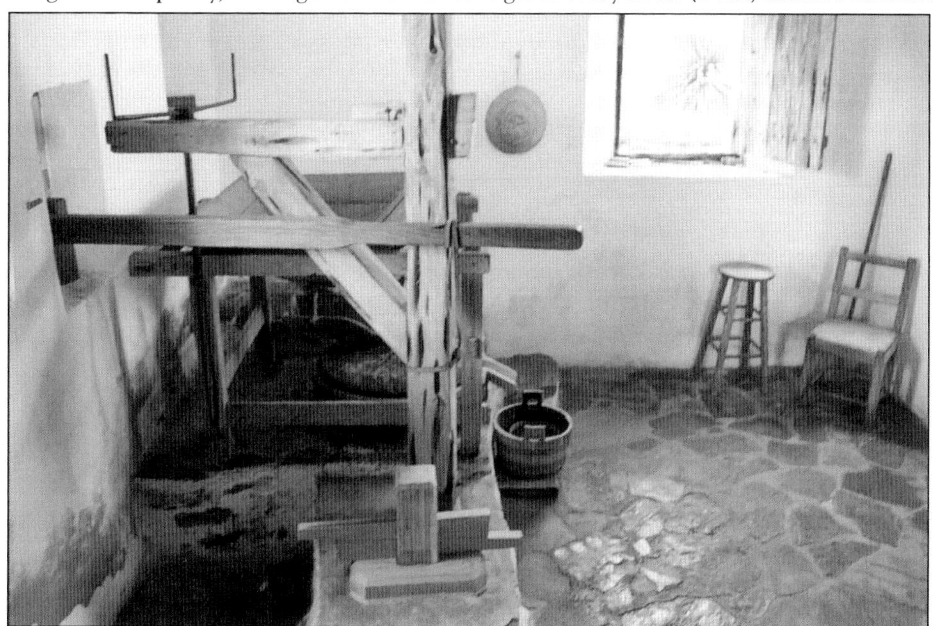

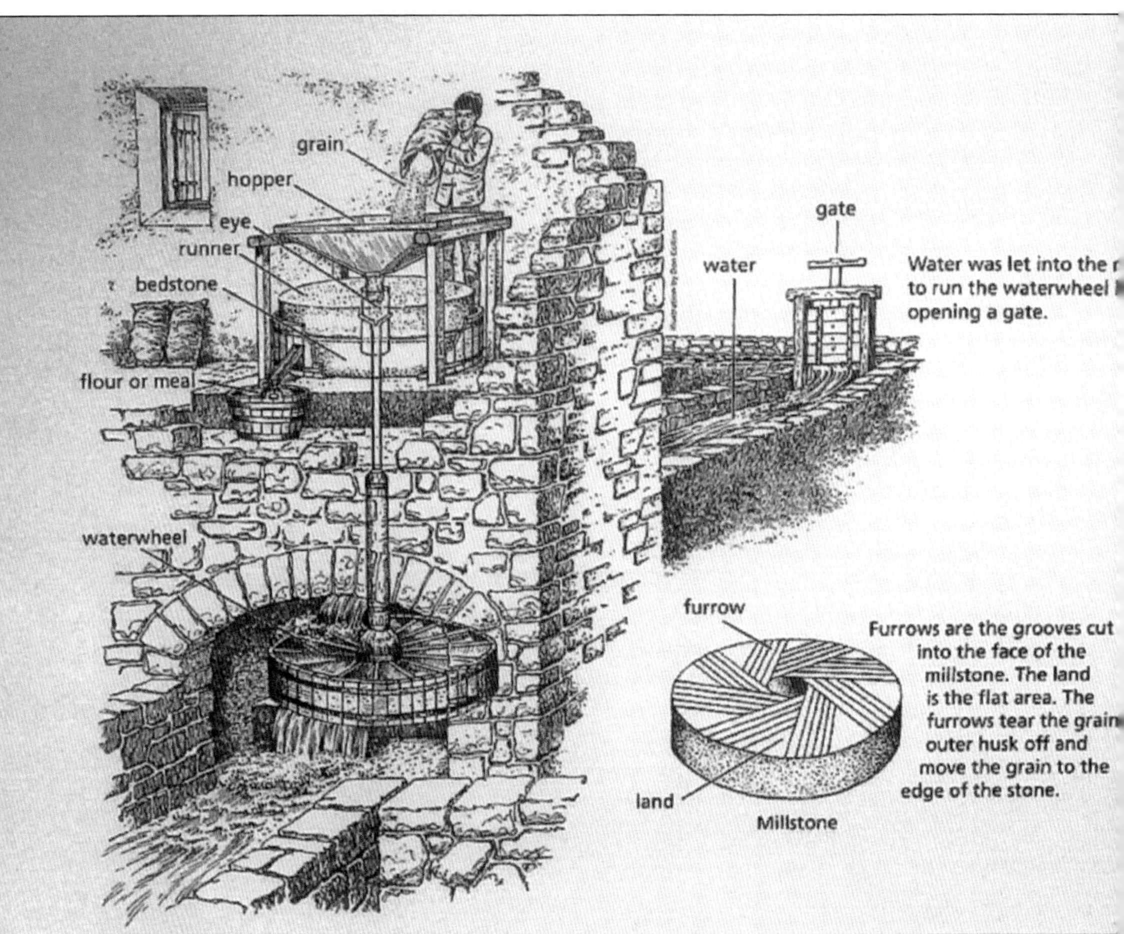

During the reconstruction of the Mission San Jose Mill in the 1930s, Pioneer Flour Mills of San Antonio provided the working parts of the mill. Ernst F. Schuchard, grandson of Carl Hilmar Guenther, who had done extensive research on the mill, led in the restoration process, doing much of the work himself. The distinguishing feature of the mill is the horizontal revolving waterwheel with a vertical shaft on which a grinding stone is mounted directly, thus eliminating any gear mechanism. This type of mill is known as a "Norse mill," and is one the simplest types of mills known. Schuchard also included reproductions of the early machinery, horizontal waterwheel, and grindstones. This photograph of an interruptive plaque at the Alamo illustrates the workings of the San Jose Mission gristmill. (Author's collection.)

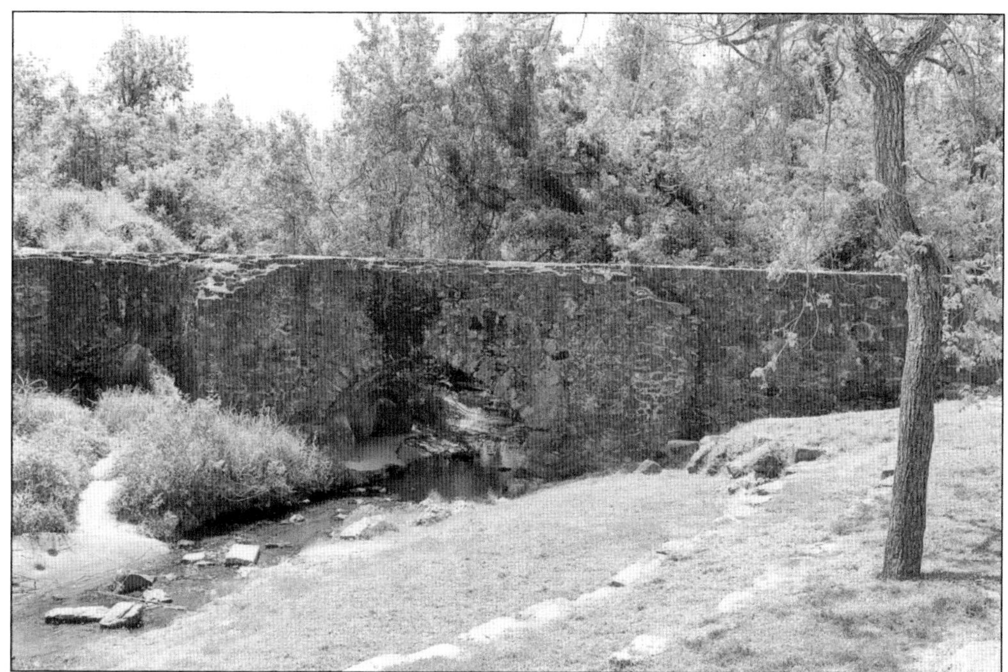

To assure a constant flow of water to the gristmills and the missions farmlands, the Franciscans and Indians dug ditches following the contours of the river valley. When the natural lay of the land imposed a blockade, they built an aqueduct. One of those aqueducts is at Mission Espada. The terrain between the San Antonio River and Mission Espada is not level and is crossed by Sixmile Creek. A Roman-style aqueduct, completed in 1745, carries the water over the creek. The Espada Aqueduct, shown in both photographs, is made of brush, gravel, and rocks. Lime salts in the water gradually cemented the layers together. This Spanish colonial irrigation system is the only structure of its type in the United States. It is now under the protection of the US Department of Interior and is a Registered National Historic Landmark. (Both, author's collection.)

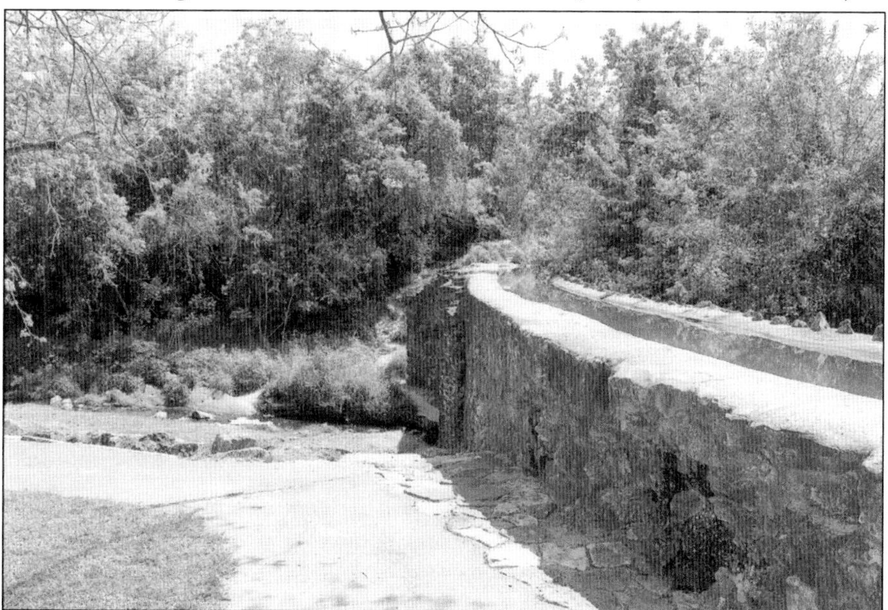

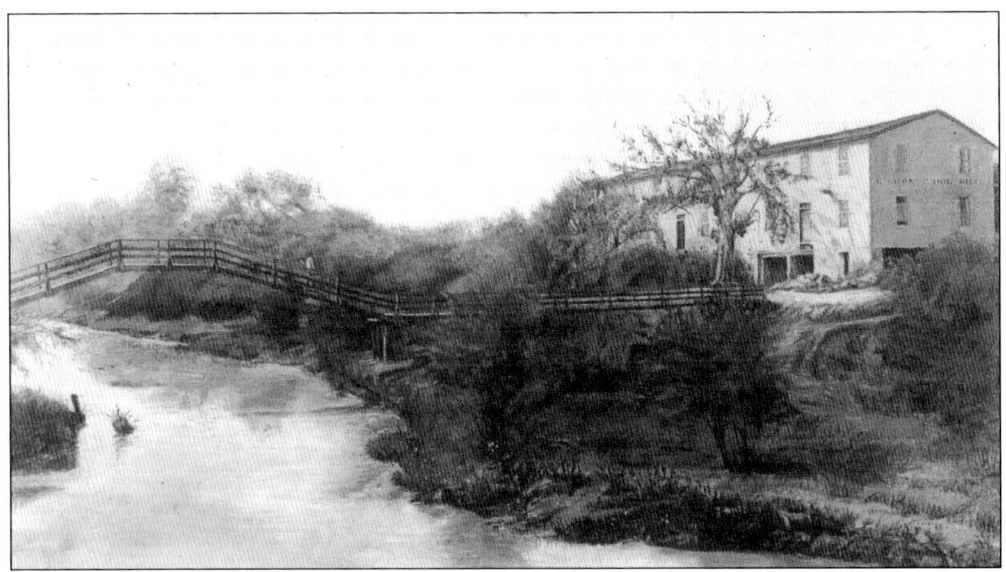

In 1879, brothers Henry and Louis Berg built the four-story General Commission, Wood and Hide Merchants mill building near Mission San Juan for wool burring, scouring, and pulling. The mill soon became the nucleus of the small community known as Berg's Mill. The mill operated until the 1920s. This oil painting of Berg's Mill is by Louise Fretelliere. (The University of Texas at San Antonio Libraries, Special Collections.)

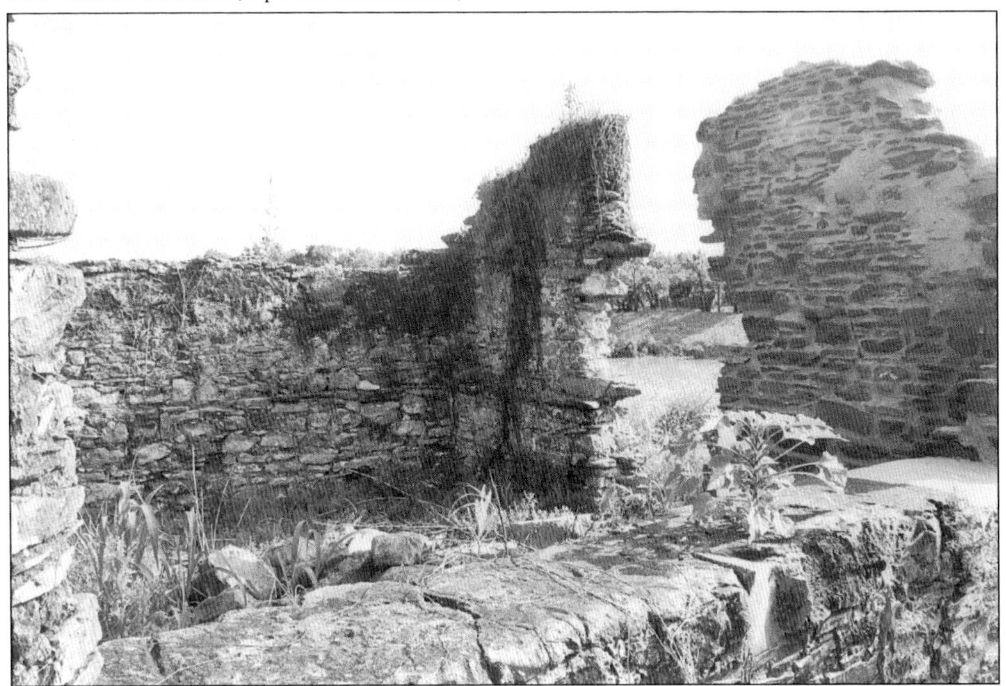

At one time, Berg's Mill was a stop on the San Antonio and Aransas Pass Railroad. A post office named Helleman opened in 1887 but closed after only three years. A second post office opened in 1892, and the names Helleman and Berg's Mill were used interchangeably. The stone ruins shown in this photograph are all that remain of the once productive Berg's Mill community. (David Carson.)

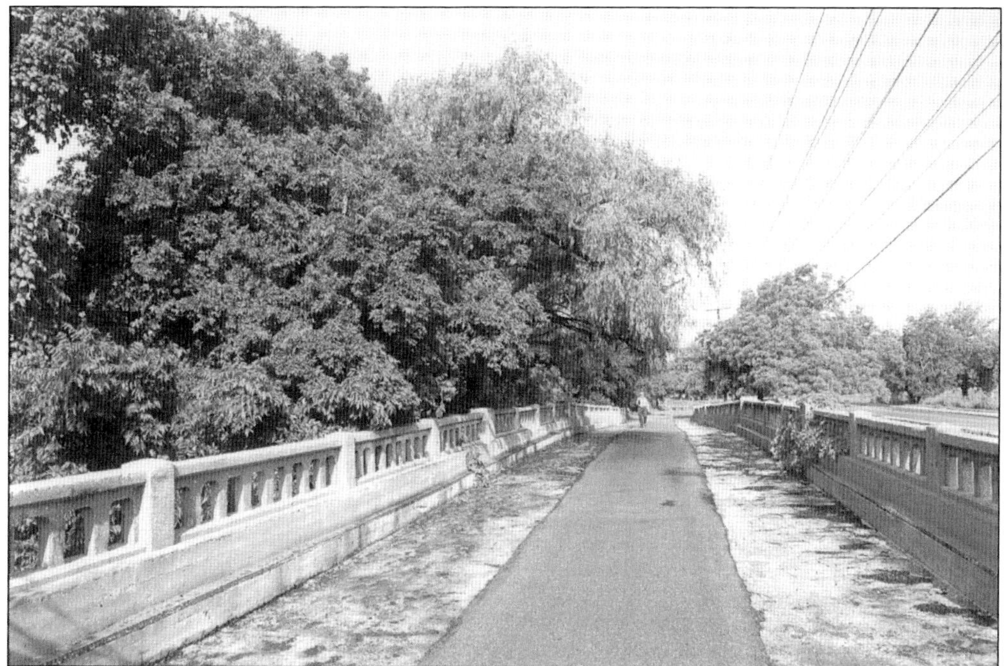

Prior to the construction of a bridge, farmers and visitors to the missions forded the river at Berg's Mill. In the 1880s, a small bridge was built across the millrace to provide access to the mills and the nearby river. This photograph shows the concrete structure that was built in 1914 to replace the 1880s bridge. In the 1960s, a new bridge was built, and this bridge was preserved for pedestrian use. (David Carson.)

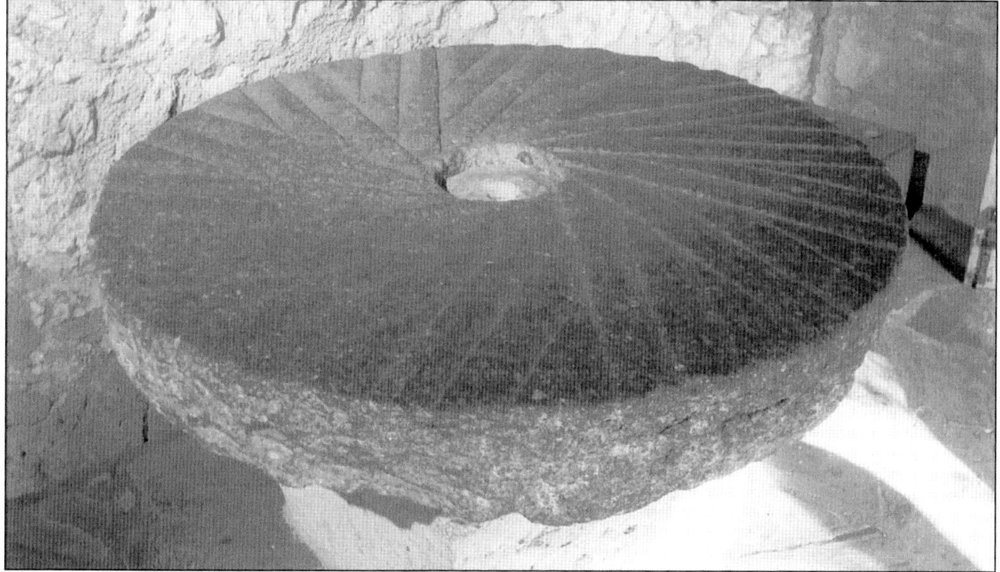

Prior to Santa Anna's attack on the Alamo in 1836, it was under siege from Gen. Martin Perfecto de Cos and his Mexican forces. Col. Ben Milam asked for volunteers to meet at the old Molino Blanco gristmill and plan a strategy to retake the Alamo. The millstone shown above is from the Molino Blanco and is now resting at the Alamo, near the spot where the stone originally ground grain. The mill operated from 1733 through 1830. (Author's collection.)

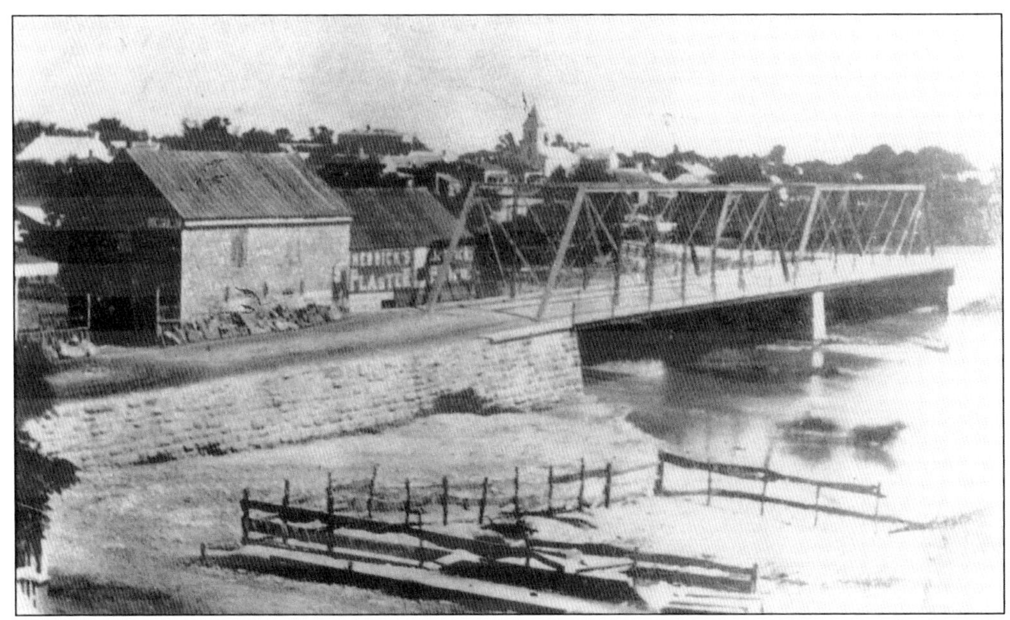

Nathaniel (Nat) Lewis selected the spot near the old abandoned Molino Blanco site for his San Antonio mill. The millstones were a gift from the king of Spain. The mill was originally run by waterpower, then it was powered by steam. The mill was in operation from 1849 to 1890. This photograph shows the mill, the Navarro Street Bridge, and St. John's Lutheran Church in the background. (Texana/Genealogy Department, San Antonio Public Libraries.)

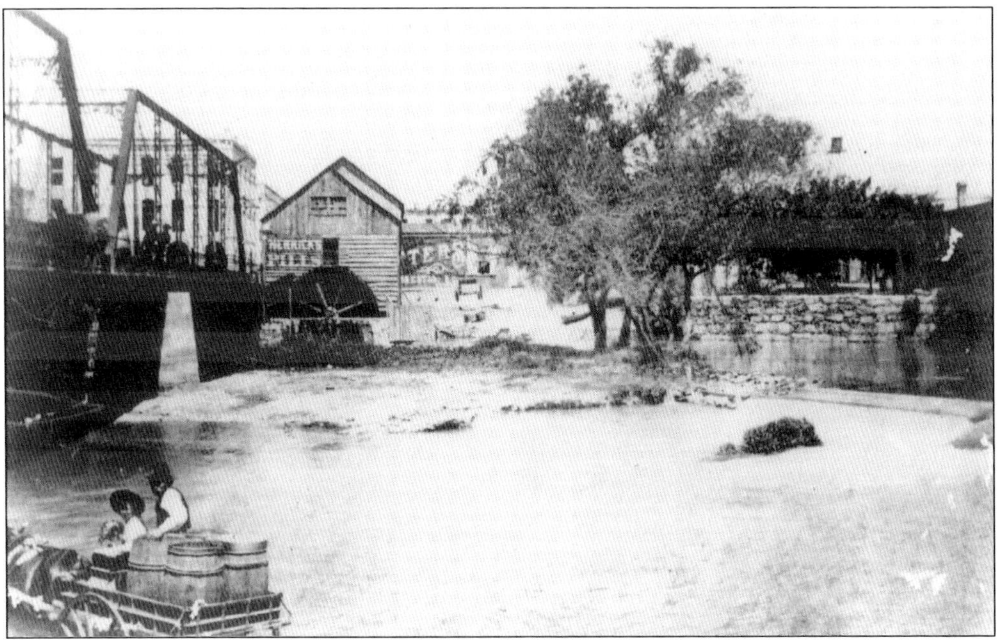

The Old Mill Bridge on Navarro Street is the subject of this photograph. Even with the bridge, wagon drivers sometimes drove their wagons through the ford of the river. Drivers would stop to soak the wagon wheels, thereby expanding the wood tight against the metal outer rim. In addition to milling, Lewis was a merchant, banker, owner of a newspaper, and manager of a wagon train. (Texana/Genealogy Department, San Antonio Public Libraries.)

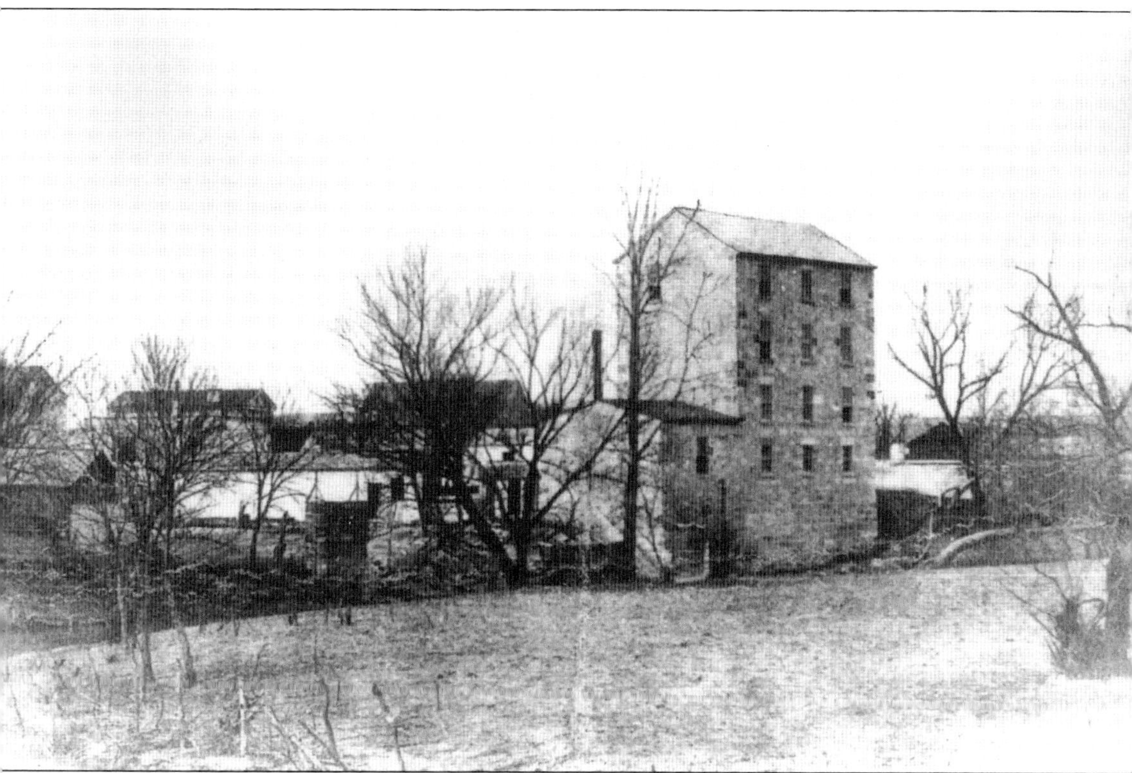

Jacob Laux, a German immigrant, came to Texas in 1844 and eventually settled in Salado. Later, after purchasing property in San Antonio, Laux moved to San Antonio where he constructed a flour mill around 1866. The mill was located on the west side of the San Antonio River, north of Travis Street and east of Soledad Street. It was a five-story stone structure and had machinery that could be driven by either waterpower or steam. It had the capacity to grind 12 to 14 bushels of flour per hour. Jacob and his wife, the former Susanne Christina Catherina Ackermann, and their family, lived on the upper floor of the mill. Laux operated the mill until 1873, after which it was leased by several different individuals. By 1883, the mill was no longer in operation. Laux died in June 1888 at age 72. Christina Laux continued to live in the mill until her death in May 1913 at the age of 84. (The University of Texas at San Antonio Libraries, Special Collections.)

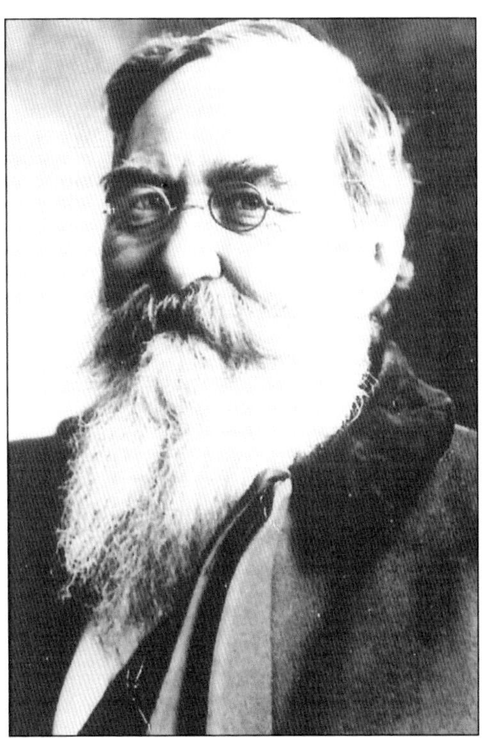

Carl Hilmar Guenther (GEN-thur) left his native Germany in 1848 seeking greater political freedom and economic opportunity in America. Guenther was trained as a millwright, so it was only natural that he built and operated a mill after he settled in Fredericksburg, Texas, in 1851. He operated this mill for several years before moving his milling operation to San Antonio. (C.H. Guenther & Son, Inc., Archives, all rights reserved.)

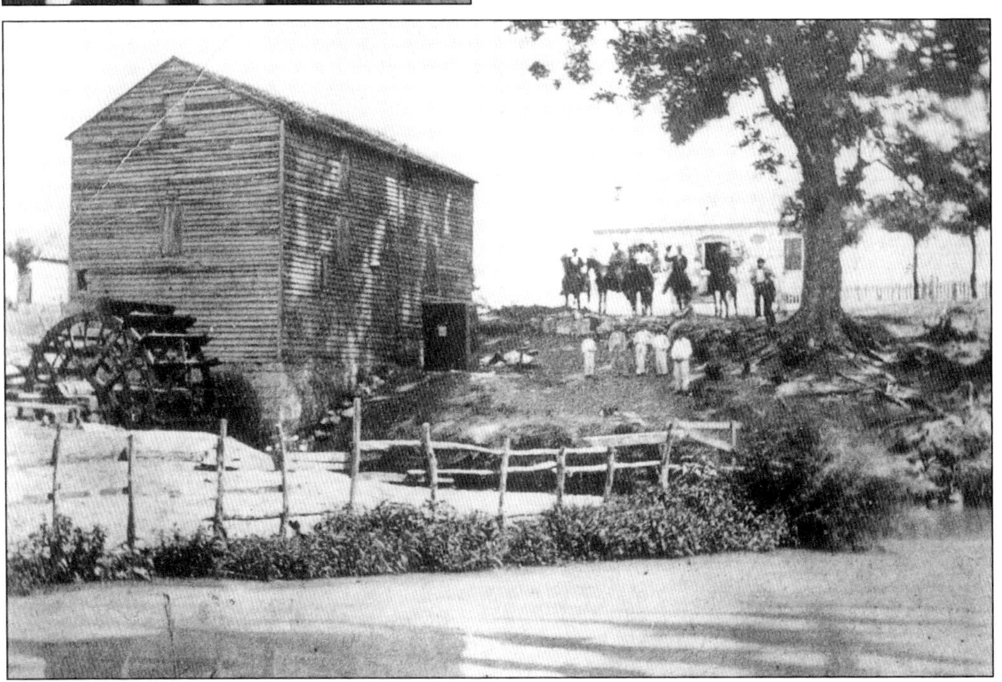

Guenther started his milling operation on the San Antonio River in 1859. His first mill, seen here, was powered by a large wooden waterwheel and was capable of grinding flour for local farmers and other consumers. During the Civil War, as a miller, Guenther was exempted from military service. He was required, however, to furnish flour to the Confederate army stationed in San Antonio. (C.H. Guenther & Son, Inc., Archives, all rights reserved.)

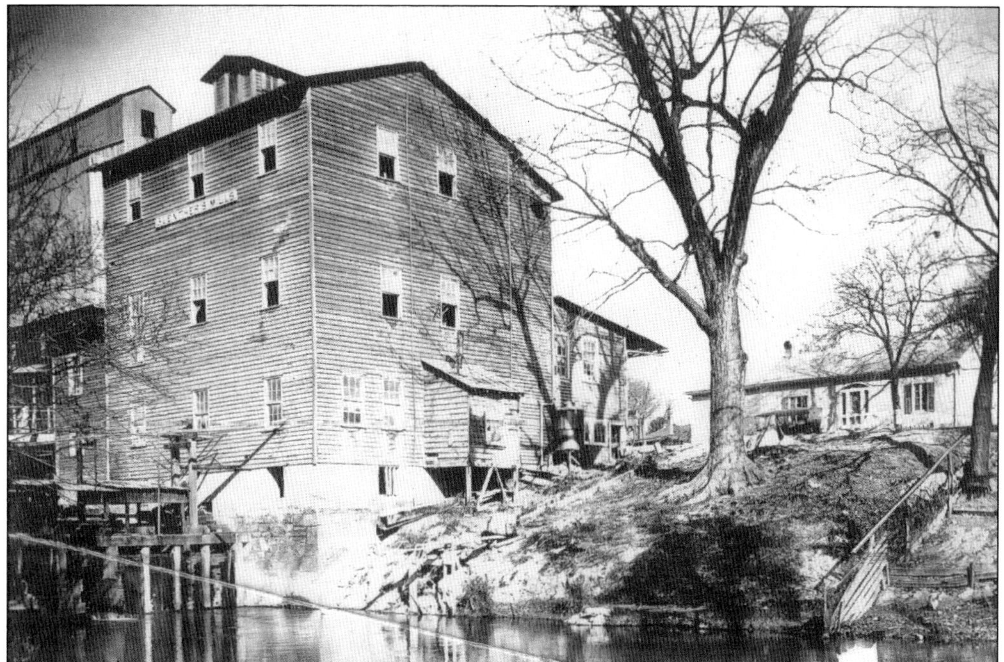

In 1877, Guenther replaced his 1859 Lower Mill with a larger, more modern mill on the same site. The new mill, seen here, used water-powered turbines in place of the old wooden waterwheel. The new turbine wheel was five feet high and six feet in diameter and was enclosed in a casing. Four pairs of millstones were installed, which increased the mill's capacity. (C.H. Guenther & Son, Inc., Archives, all rights reserved.)

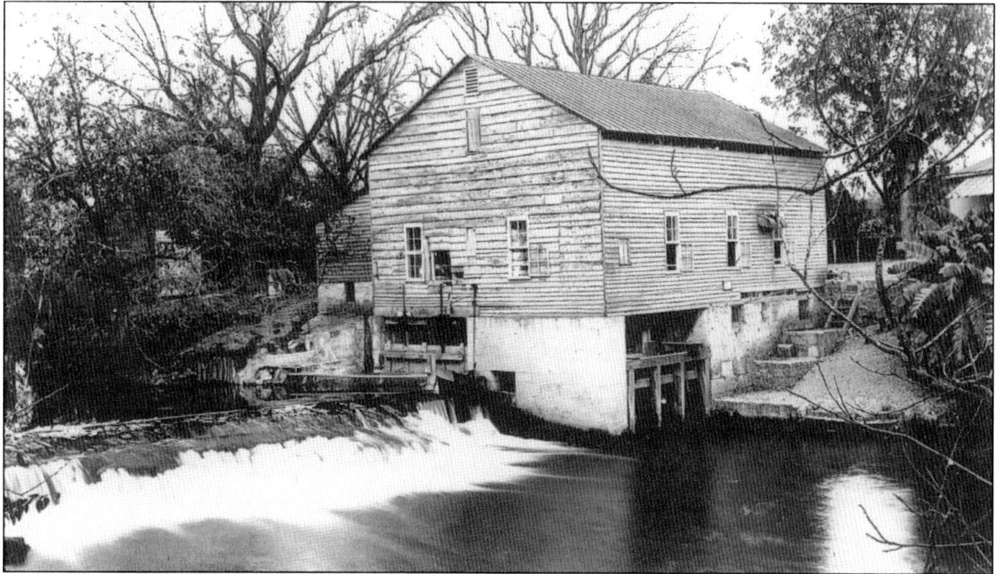

In 1868, Guenther built another dam and mill upstream, closer to town on Washington Street across from the US arsenal. This Upper Mill became the primary retail outlet for both mills until about 1900, when operations began exclusively out of the Lower Mill. The vacated Upper Mill later housed a macaroni factory, an ice cream company, and a garage. (C.H. Guenther & Son, Inc., Archives, all rights reserved.)

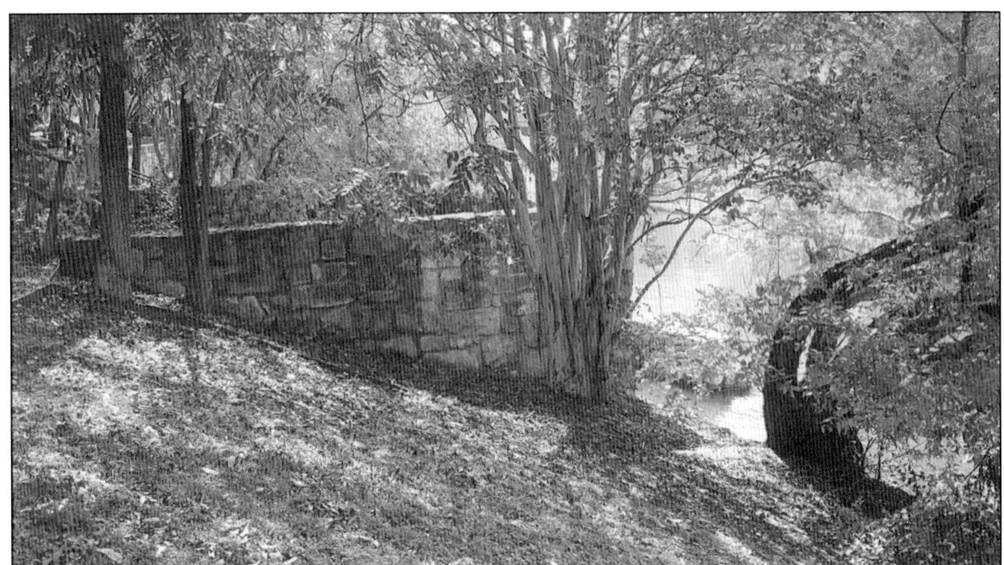

Guenther's Upper Mill building was razed in 1926 as part of flood prevention measures along the San Antonio River. The foundation and millwheel pit were left in place but were removed later, as they remained an obstruction in the channel. The mill foundation was rebuilt when the Guenther Mill Park was created as part of the San Antonio River Channel Improvements Project in the early 1980s. This photograph shows the rebuilt foundation. (Author's collection.)

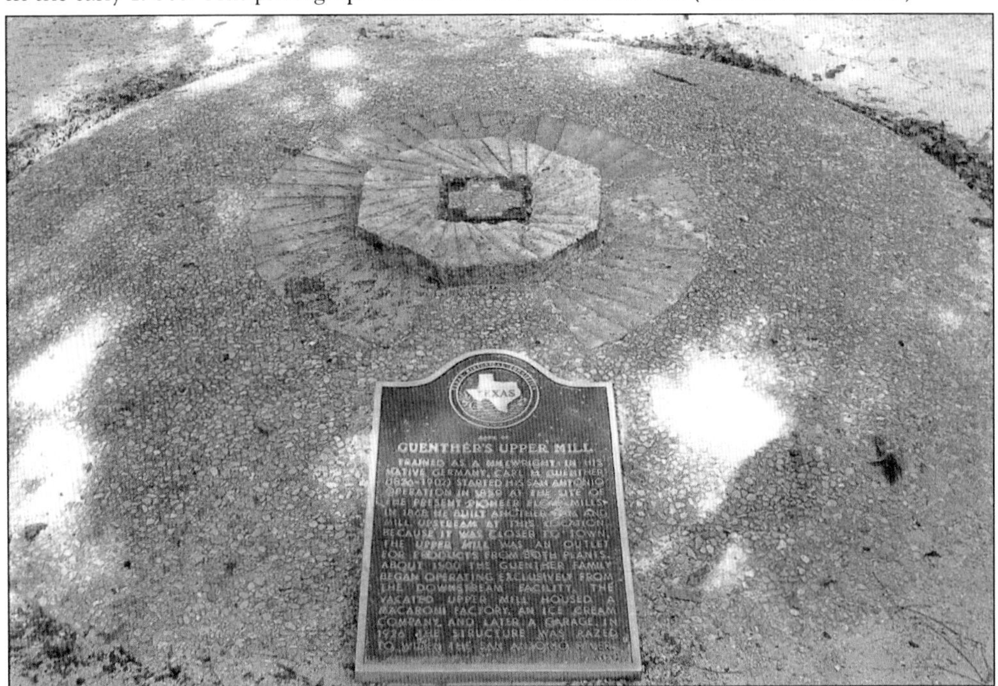

Throughout the years, Guenther built four new mills to care for the ever-expanding needs of San Antonio and beyond. This photograph shows an almost-complete sectional millstone embedded in the area near the Upper Mill ruins. This stone and the Texas Historical Commission marker, located in Guenther Mill Park at the San Antonio River Walk and Arsenal Street, are in recognition of Guenther's contributions to the milling industry. (Author's collection.)

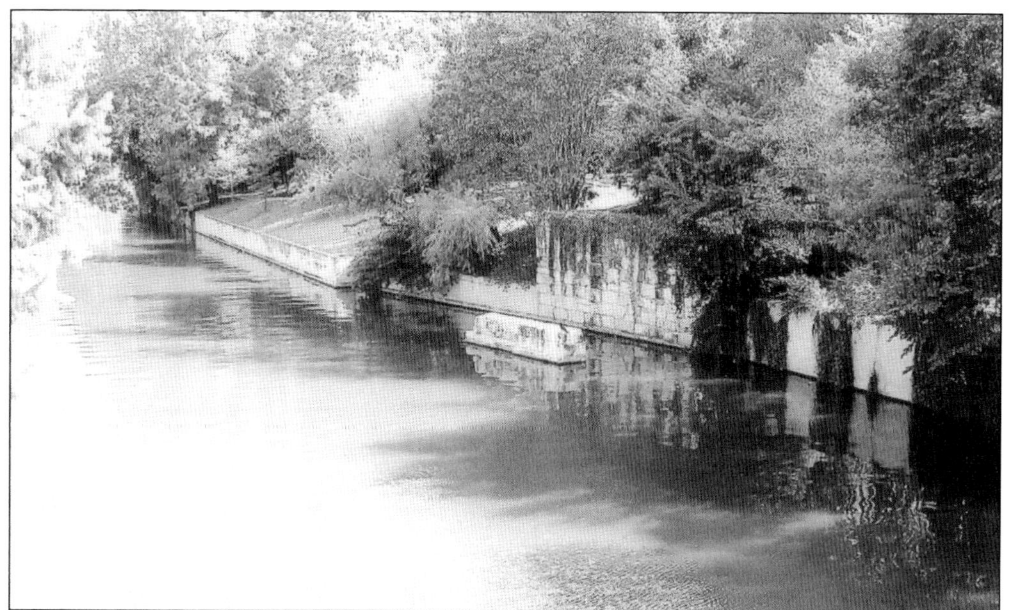

This photograph offers a view of the vine-covered, rebuilt foundation of Guenther's Upper Mill. The limestone block in the water is a part of the original millwheel pit, which can be seen in the previous photograph of the Upper Mill. This view is looking north from the Arsenal Street Bridge crossing the San Antonio River. (Author's collection.)

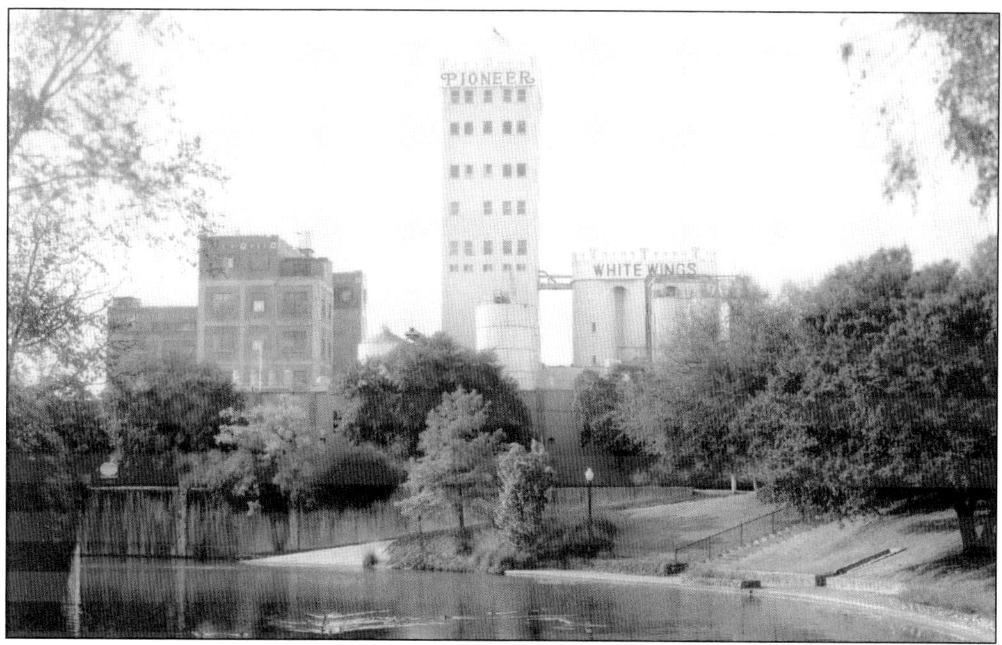

In 1898, Guenther's milling business was incorporated under the name C.H. Guenther & Son, Inc., using the trade name Pioneer Flour Mills, the name by which it is recognized today. Pioneer continues to produce retail and foodservice grain products for distribution throughout the United States. It is the oldest family-owned business in Texas and the oldest continuously operated family-owned milling company in the United States. (C.H. Guenther & Son, Inc., Archives, all rights reserved.)

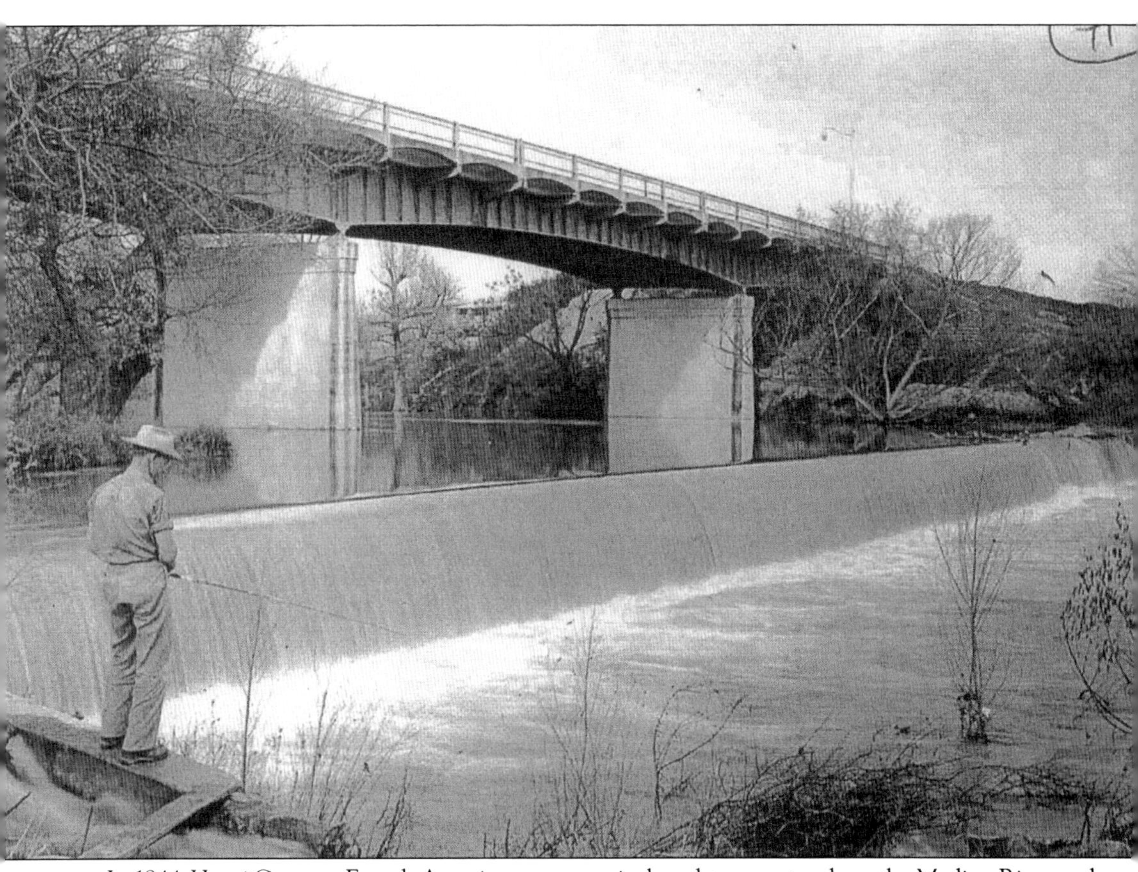

In 1844, Henri Castro, a French American empresario, bought property where the Medina River and the Old San Antonio–El Paso Road met. Castro then founded the Alsatian colony of Castroville in 1846. In 1849, Cesar Monod, a Castroville colonist, purchased an original town lot from Castro and then purchased the adjacent lot from Michel Simon. Monod built a one-story structure that he used as a residence and a dry goods store. In 1853, John and Rowena Vance purchased the Monod property. It became known as the Vance Hotel when they added a second floor to the residence and began renting rooms to travelers. In 1854, Vance sold the lower, riverfront portion of the property to Laurent Quintle and George Louis Haass, who in turn built a stone-and-wood dam across the Medina River, a two-story gristmill, and a stone-lined underground millrace running from the dam to the mill. (Landmark Inn State Historic Site, Texas Historical Commission.)

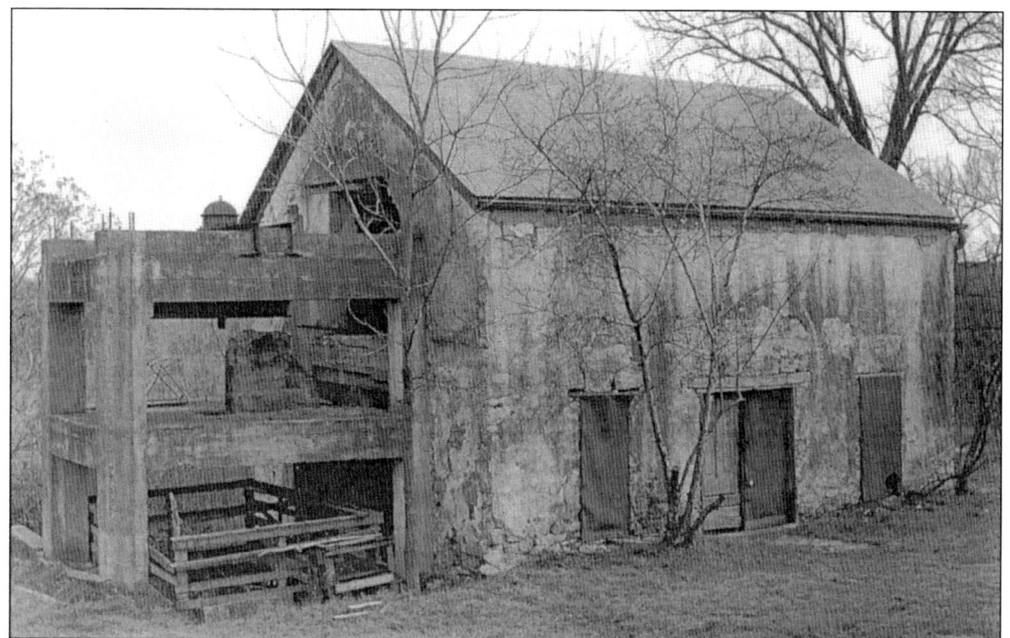

The Quintle and Haass Mill consisted of a basement that housed the machinery connecting the waterwheel and the millstones; a first-floor milling room where the grinding and business transactions occurred; and a second floor for storage of grains and equipment. The mill was powered by a large wooden undershot waterwheel, which was turned by the force and weight of the water flowing beneath it. (Landmark Inn State Historic Site, Texas Historical Commission.)

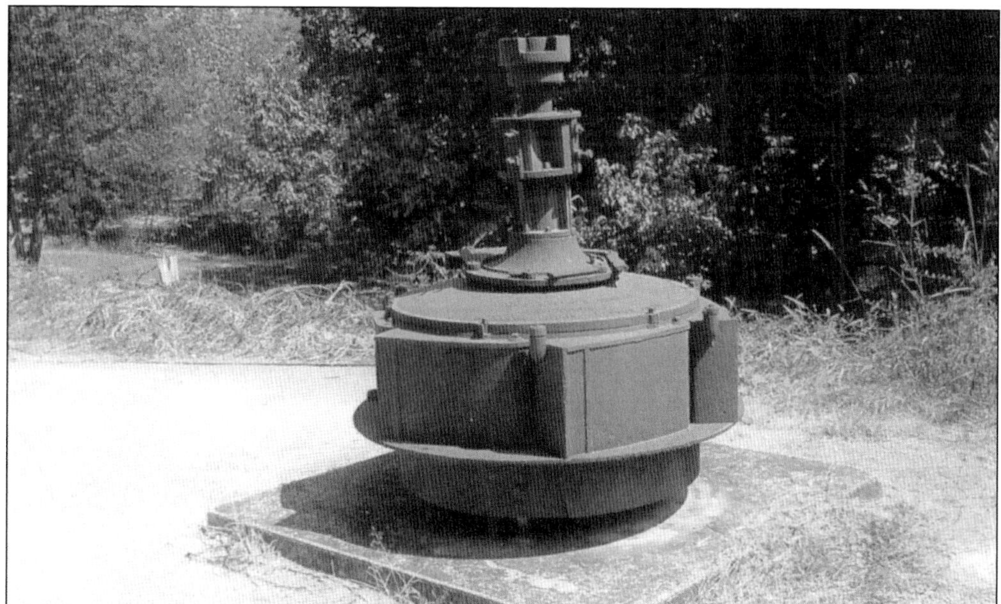

After the Civil War, the picturesque wooden waterwheel was replaced by a water turbine. A water turbine is a horizontal waterwheel enclosed in a case with adjustable gates that direct and regulate the flow of the water against the wheel. At least three different turbines were used at the mill. The first one was sold and replaced by a second turbine, which now sits on display outside the mill building. (Author's collection.)

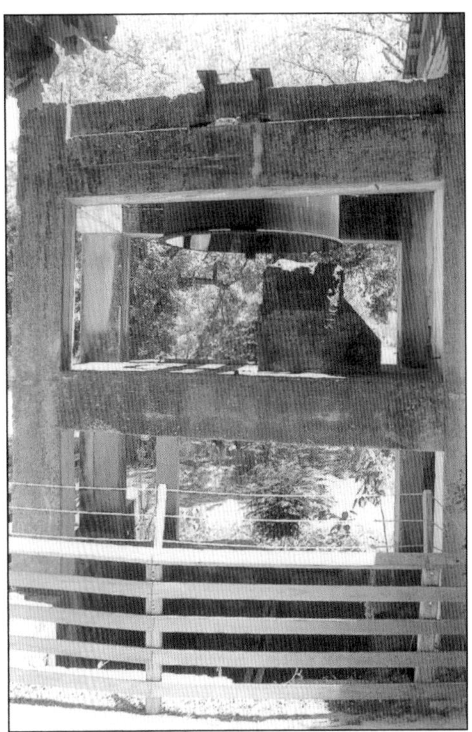

The turbine was housed in a wheel pit to the north of the mill building. The turbine itself was connected to a vertical shaft that transferred power to a horizontal wheel at the top of the shaft. A flat belt connected to a large flat pulley on the upper end of the shaft transferred power into the mill at the second-floor level. (Author's collection.)

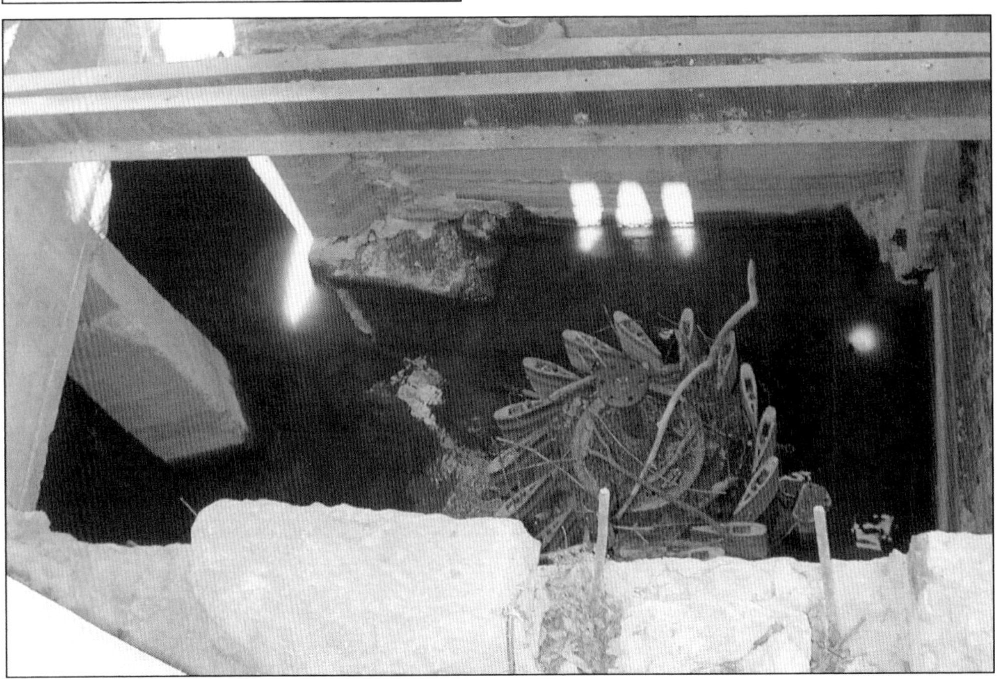

This view from inside the mill shows a section of the last turbine used. It sits at the bottom of a 12-foot-deep pit. The large opening on the left is where water entered the mill pit through the underground millrace from the Medina River dam, 400 feet upstream. The water exited the mill pit through an opening to the right, not shown, and rejoined the river downstream via the tailrace. (Author's collection.)

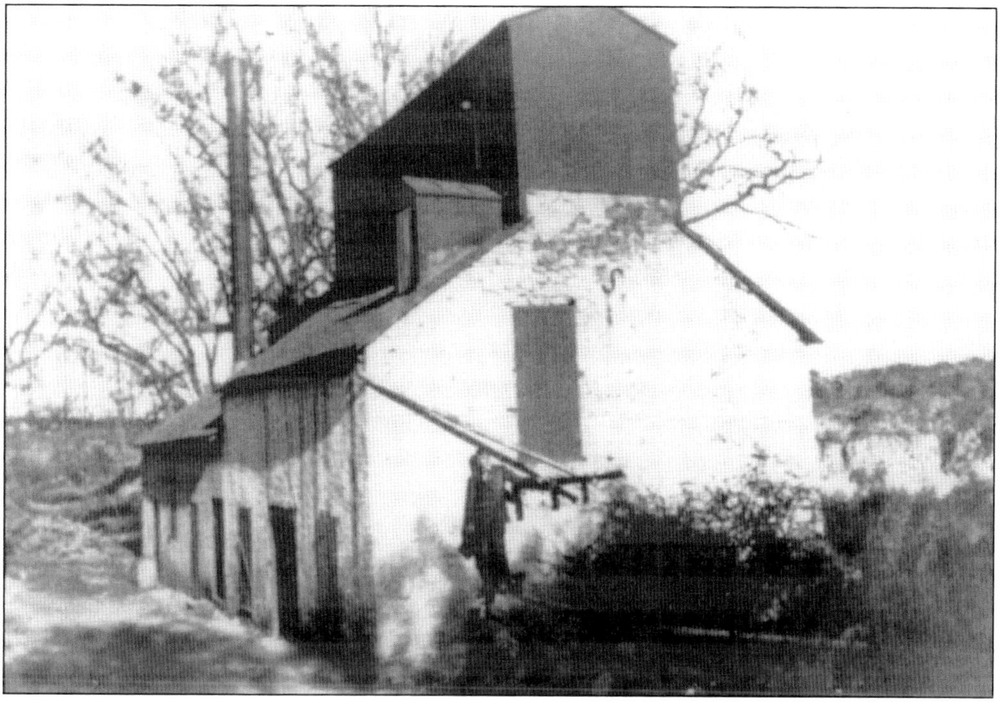

In 1876, Joseph Courand Sr. bought the mill and operated it as a steam- and water-powered flour, grist, and lumber mill. It was operated as a custom mill where the farmer's grain was ground and the meal or flour returned to him after the miller took out his toll. The miller's toll was commonly set by tradition. (Landmark Inn State Historic Site, Texas Historical Commission.)

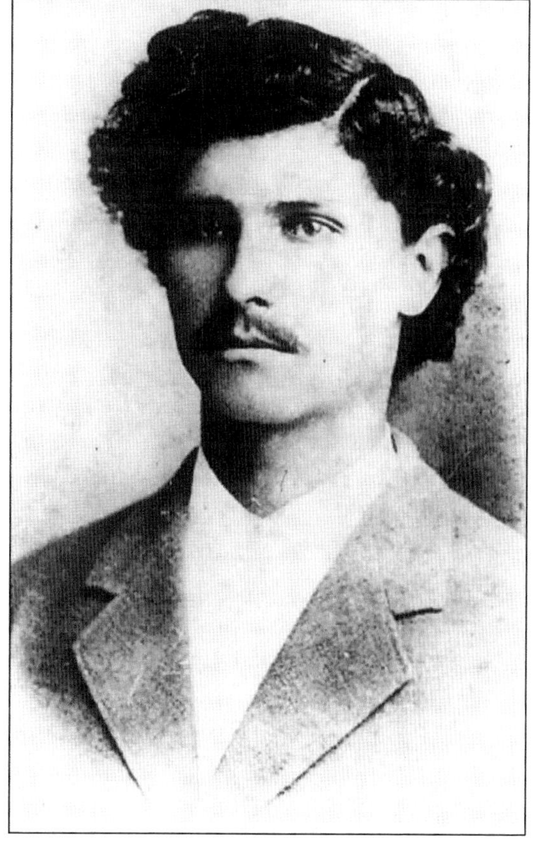

After the death of his father in 1879, Joseph Courand Jr. assumed management of the mill. He modernized its equipment so that it could process corn, wheat, cotton, and lumber, making it the industrial and commercial hub of Castroville. In a commercial mill, the farmer did not necessarily receive meal from his own grain but an equivalent amount in flour or meal. (Landmark Inn State Historic Site, Texas Historical Commission.)

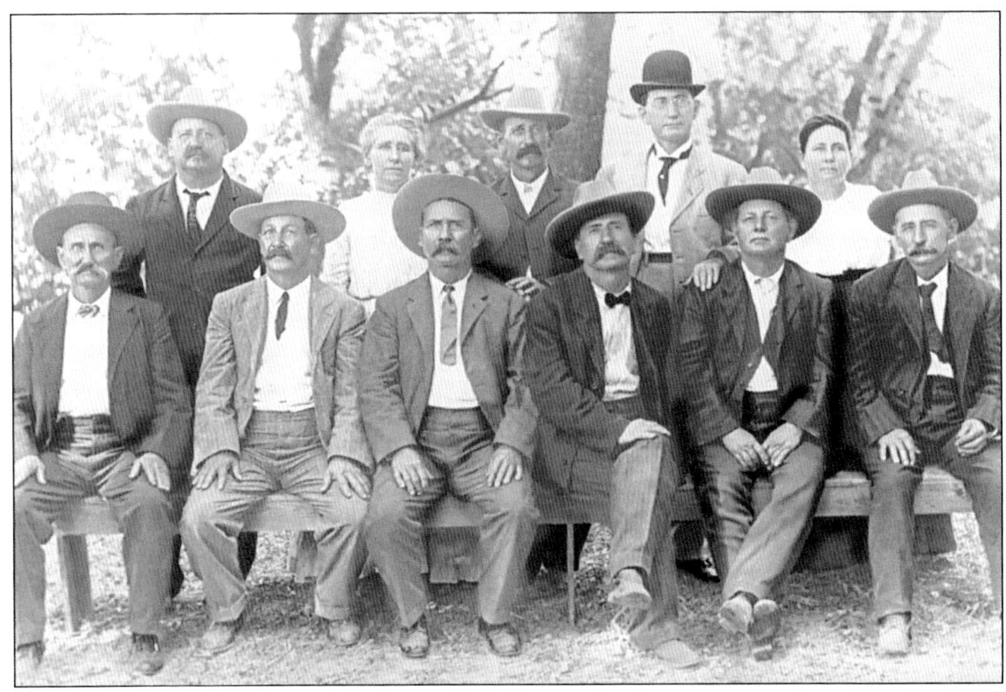

The 11 children of Nicholas and Catherine Tschirhart gather for a portrait at the mill in the 1880s. They are, from left to right, (first row) Joe, Sebastian, Edward, Leo, Henry, and Nick; (second row) Louis, Caroline, August, Emil, and Catherine. Nicholas Tschirhart and Catherine Meyer married in Castroville on February 7, 1846, and soon became active citizens of the community. Tschirhart descendants continue to hold family reunions in Castroville. (Tschirhart family collection.)

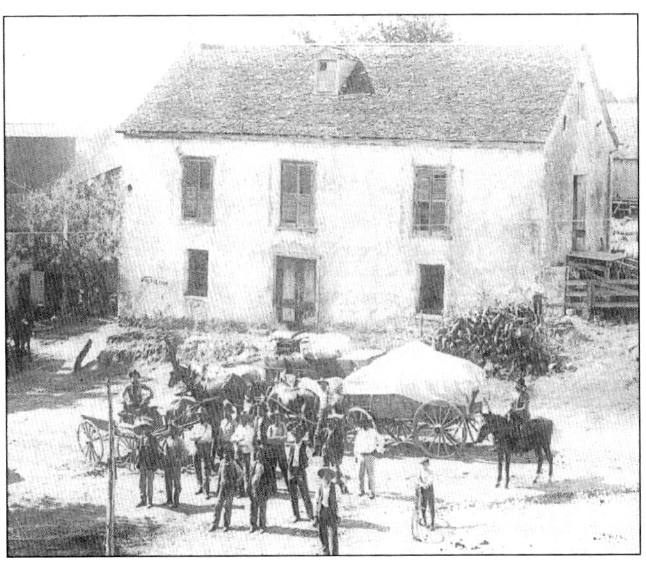

This 1900s photograph shows a busy day at the mill. Courand continued to improve and modernize the milling complex by replacing the millstones with a roller mill and by installing a 35-inch Leffel turbine waterwheel. In 1925, Jordan T. Lawler bought the entire complex, and the milling operation ceased. Lawler converted the mill to a hydroelectric power plant, which provided Castroville with electricity until 1935. (Landmark Inn State Historic Site, Texas Historical Commission.)

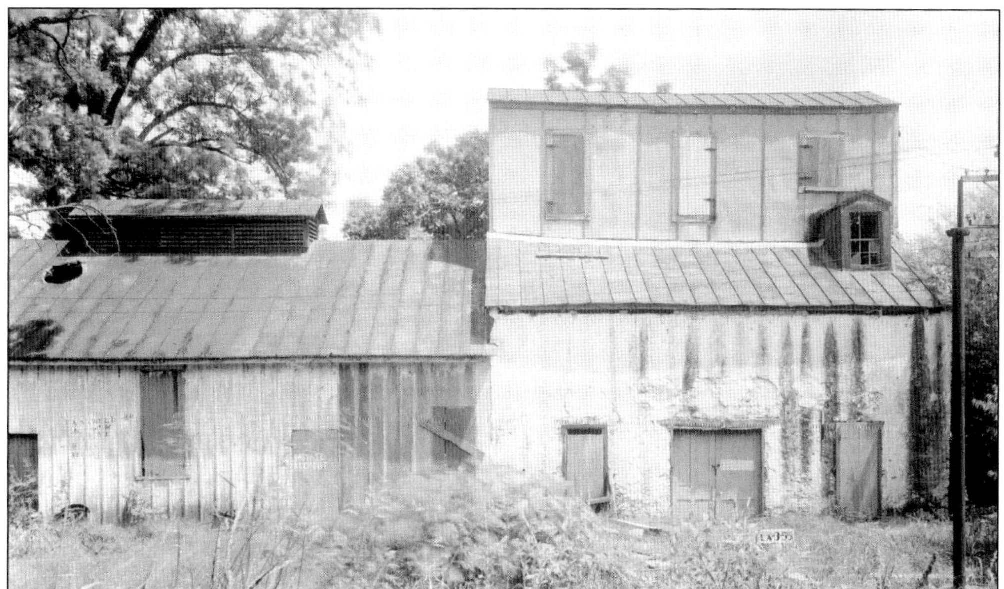

This photograph shows the west (front) elevation of the Quintle and Haass Mill as it appeared in 1936. The purpose of the elevated section of the mill roof was probably to provide room so an elevator could be added. The elevator was an endless belt or chain with buckets attached that delivered grain into an elevated hopper. (Library of Congress, Prints & Photographs Division, Historic American Building Survey [HABS] TEX, 163-CAST,4-1.)

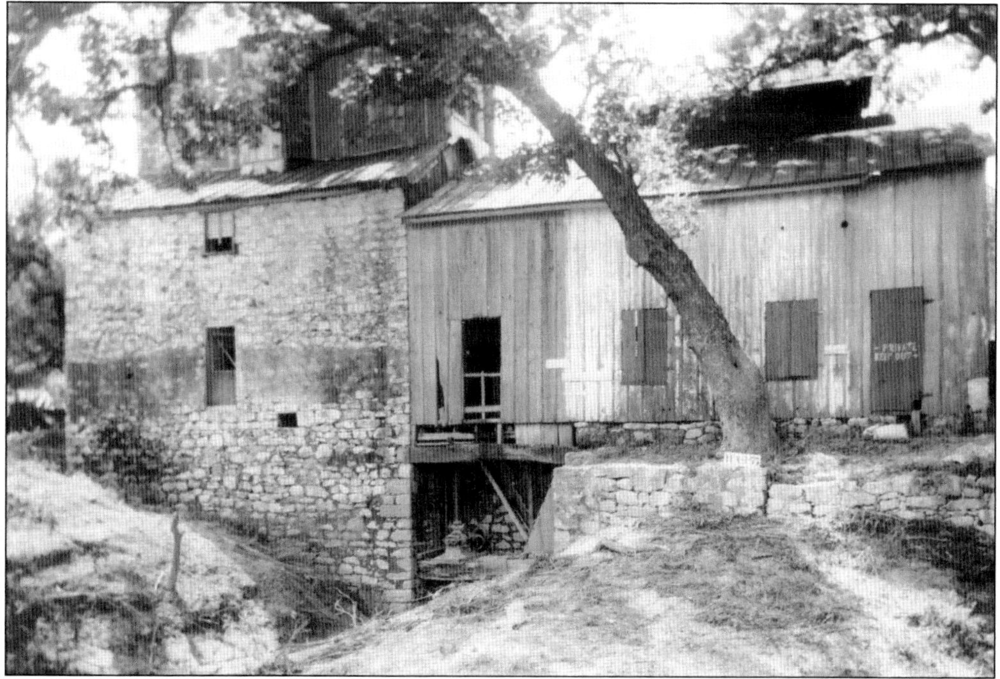

This photograph shows the east (back) elevation of the Quintle and Haass Mill as it appeared in 1936. The turbine and spillway can be seen. The frame addition was removed in the 1940s, and the stone building was reroofed later and used for storage. (Library of Congress, Prints & Photographs Division, HABS TEX, 163-CAST,4-4.)

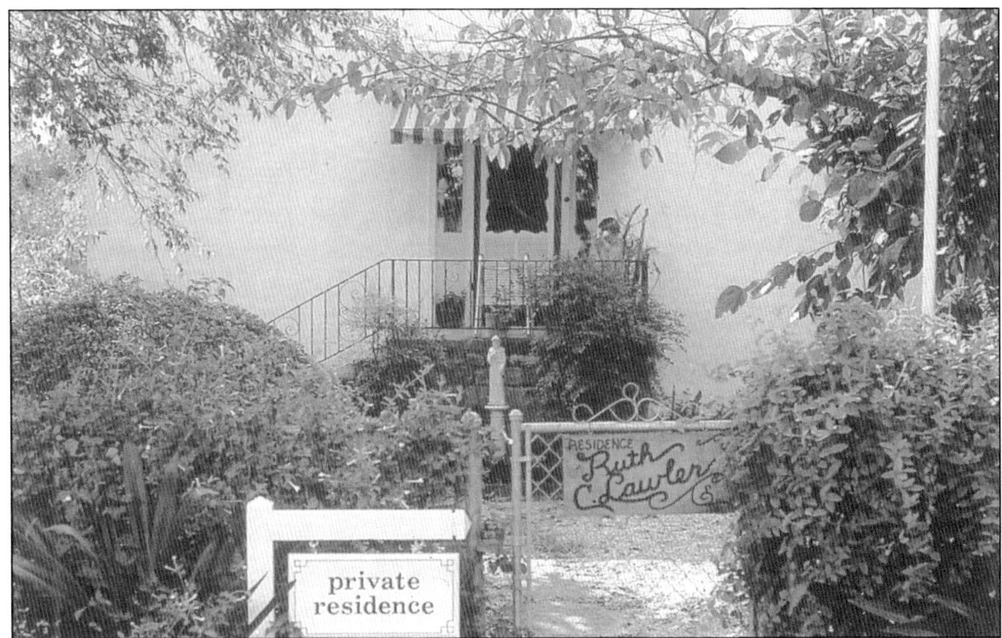

Jordan Lawler's sister, Ruth Lawler, a local schoolteacher and community leader, made the Vance residence her home. In 1942, she began renting rooms to families of servicemen stationed at the nearby Air Force base. The name was then changed to Landmark Inn. To preserve the site for the education and enjoyment of future generations, Ruth Lawler gave the property to the State of Texas in 1974. Restoration of the property began in 1975 under the supervision of the Texas Historical Commission and was completed in 1981, when the Landmark Inn, including the mill and all other property, was opened as a historic site. After a second restoration, it reopened in December 2015 as a historic site and a bed-and-breakfast. The photograph below shows the inn with its newly plastered exterior walls. (Both, Landmark Inn State Historic Site, Texas Historical Commission.)

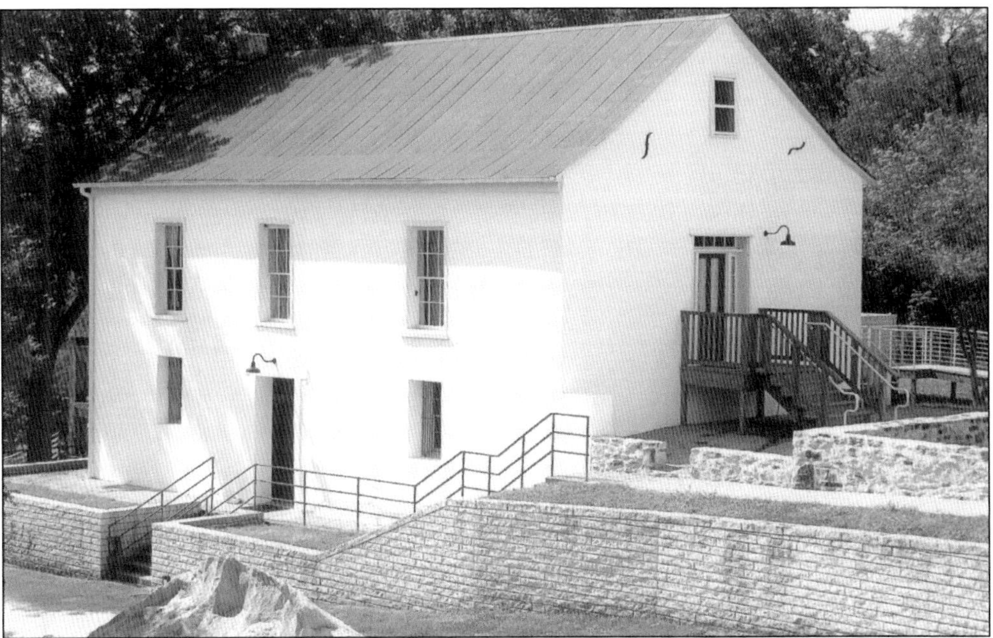

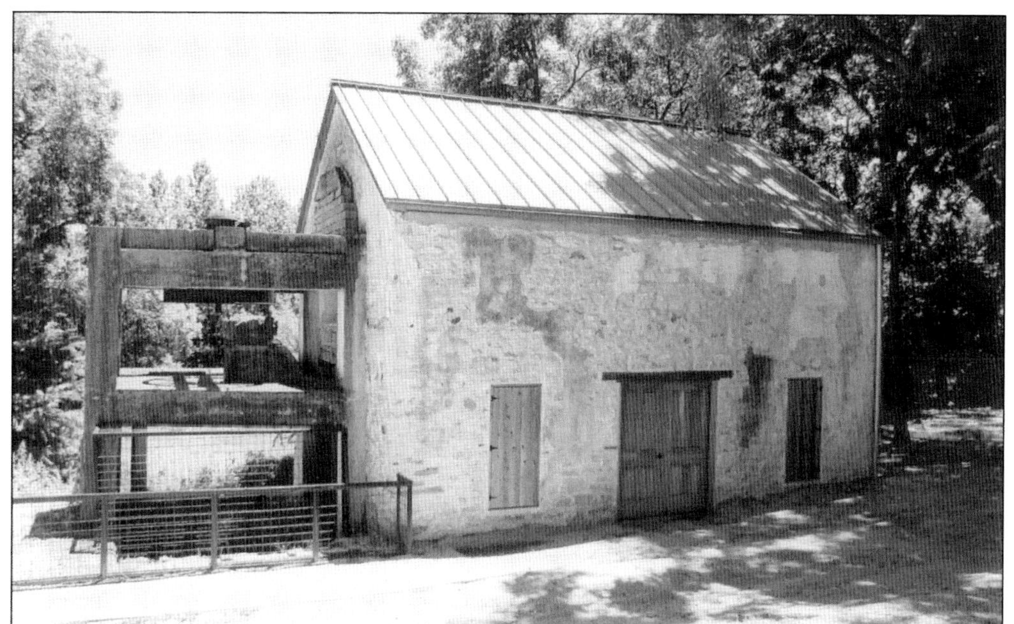

The refurbished 1854 Quintile and Haass Mill will be used for educational purposes and will be included in the site's living history exhibits. Visitors will be able to experience the mill as it was when a notice was published in the February 20, 1854, edition of the German-language newspaper *San Antonio Zeitung* that stated, "*Da wir in Castroville eine Muhle und Cotton-Gin.*" (Landmark Inn State Historic Site, Texas Historical Commission.)

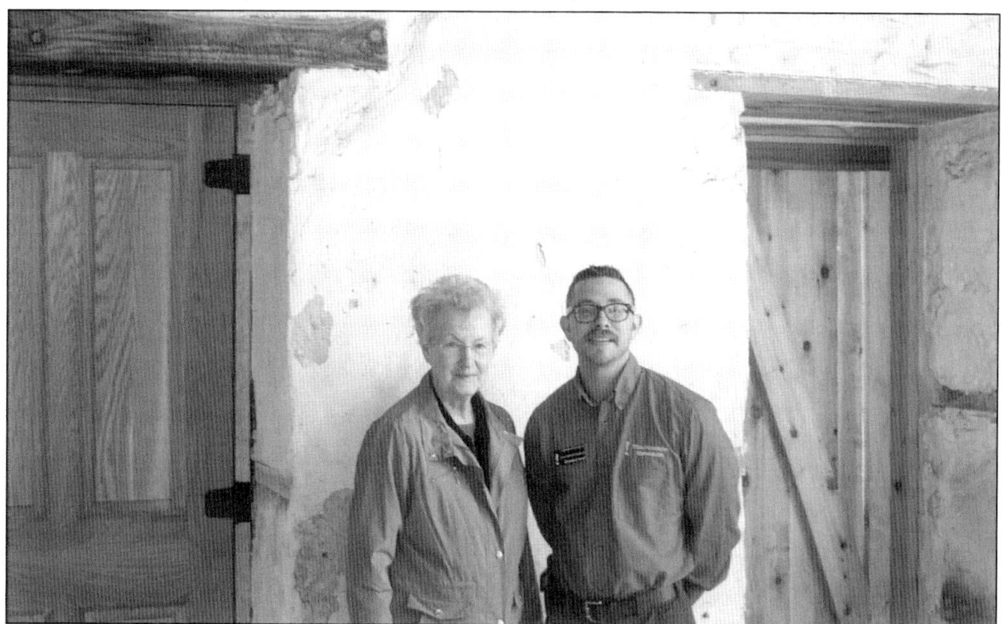

Author Charlene Carson stands next to Brandon Aniol inside the freshly renewed millhouse. Aniol, site educator of the Landmark Inn, explained that the next phase of renovation is to restore the turbine pit to its original working order. Meanwhile, the Texas Historical Commission operates the complex as a public historic site with a unique bed-and-breakfast steeped in Texas history. The site is open year-round, and visitors are always welcome. (Author's collection.)

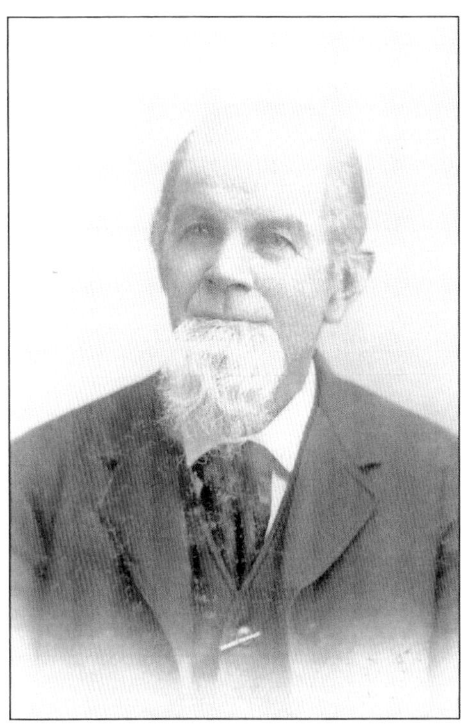

William Dietert, a millwright by trade, emigrated from Germany in 1854 and settled in Boerne, Kendall County. In 1857, Dietert built a tall frame millhouse and a low-water dam across the crystal-clear waters of Cibolo Creek near present-day State Highway 46 and South Main Street. He then built a gristmill and sawmill on the east side of the dam. A cotton gin and press were added later. (Patrick Heath Public Library, Boerne.)

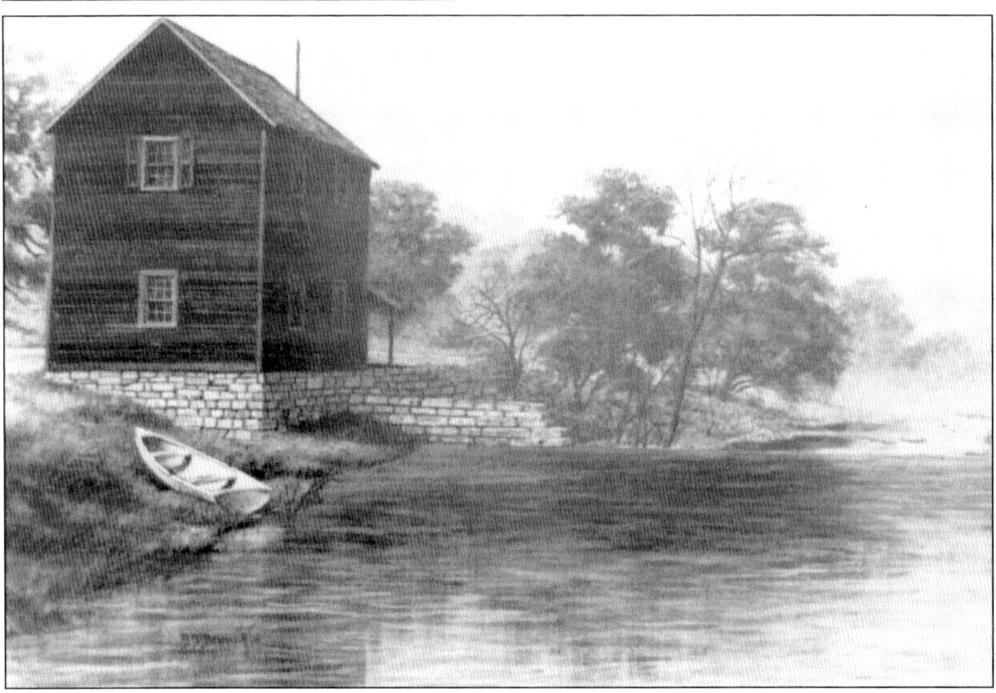

Unfortunately, there are no photographs of the Dietert Mill. However, in 1983, artist Bill Barrick of Cedar Park was commissioned to do a conceptual painting of the mill. The mill was first run by waterpower, then wood, and then steam. William's younger brother Heinrich (Henry) joined him in the milling business in 1856, and together the brothers built a strong, successful business. (Patrick Heath Public Library, Boerne.)

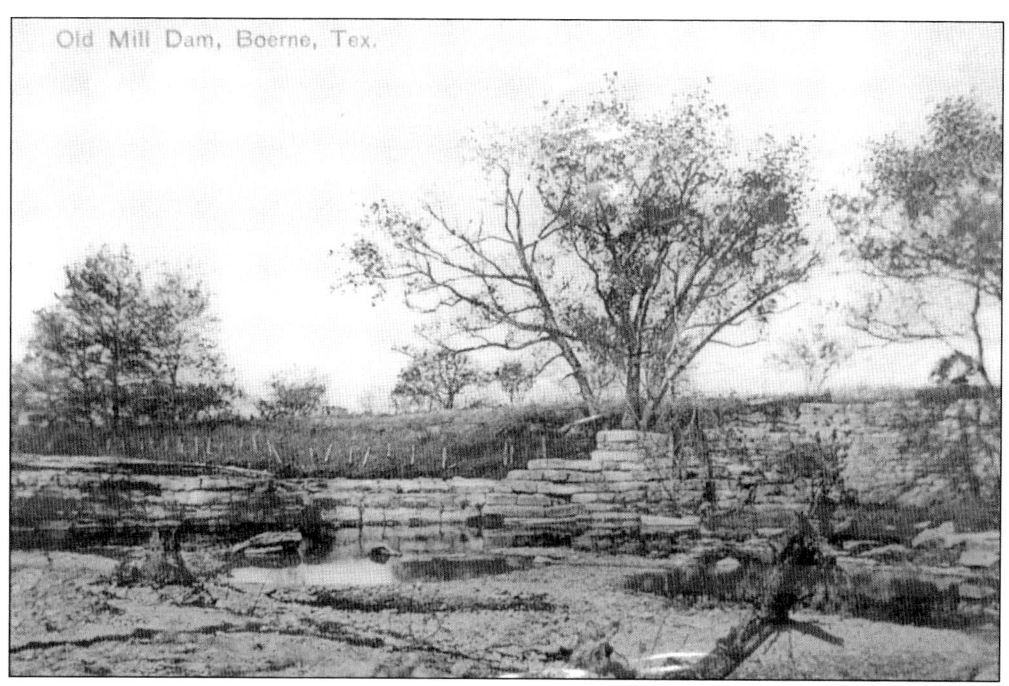

The Dietert Mill burned in the late 1870s or early 1880s, leaving only the ruins of the mill and the stone wall dam. The above c. 1915 photograph shows the location of the mill. After the mill burned, the Dietert brothers devoted themselves to agriculture and ranching. The early-1900s photograph below shows that the mill dam site became a very popular place for couples to visit while out on a Sunday afternoon buggy ride. On July 16, 1912, the William and Henry Dietert families deeded the dam site to the City of Boerne for $1. The site remains a popular park for locals and visitors alike. (Both, Patrick Heath Public Library, Boerne.)

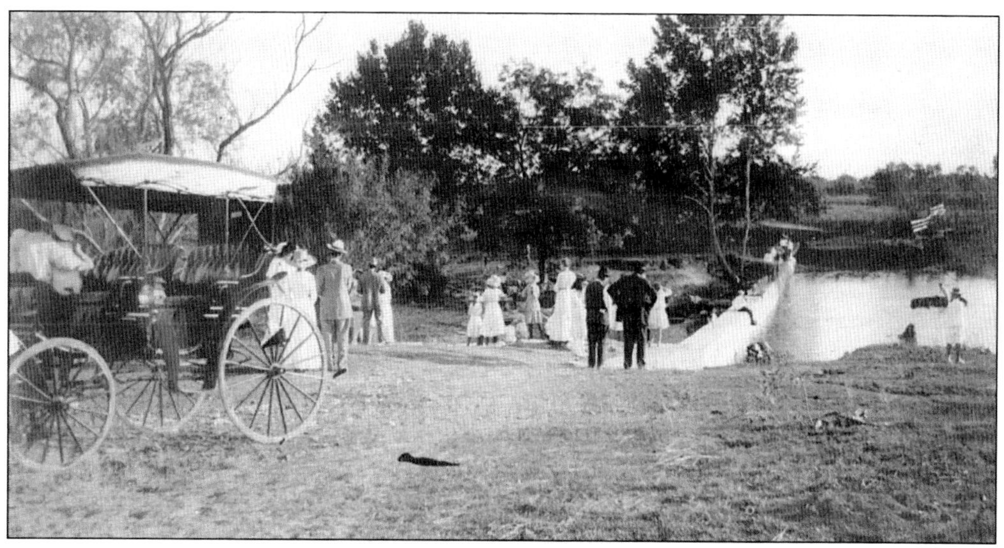

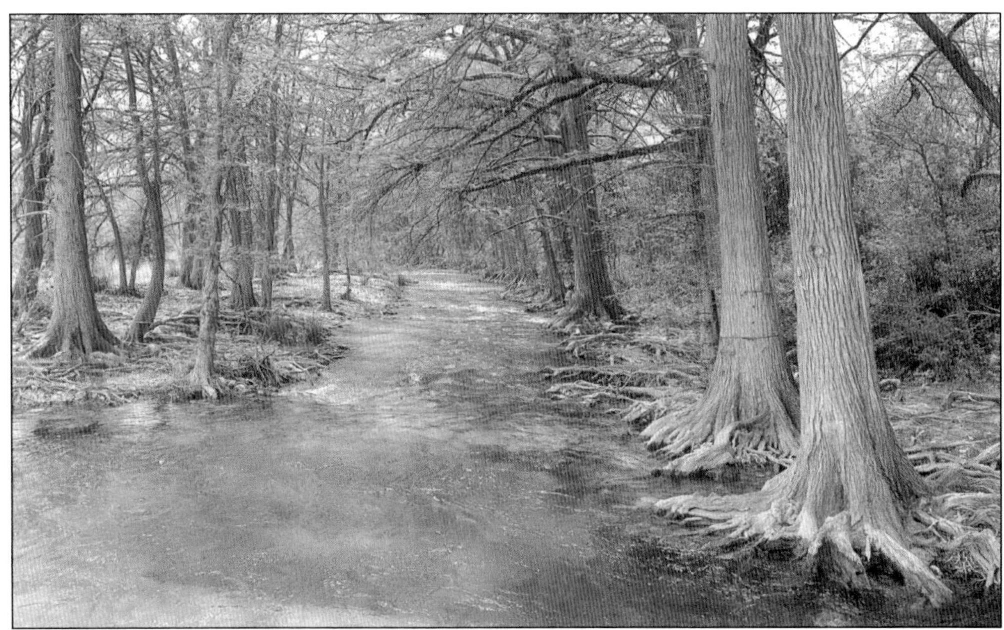

This a view of the Frio River, flanked by cypress trees, as it winds its way through Real County, home of the Rio Frio Mill and Gin. *Frio* is Spanish for "cold," referring to the spring-fed coolness of the river. After the East and West Frio Rivers join the Dry Frio River near Uvalde, the river flows generally southeast for 200 miles until it empties into the Nueces River near Three Rivers, Texas. (Author's collection.)

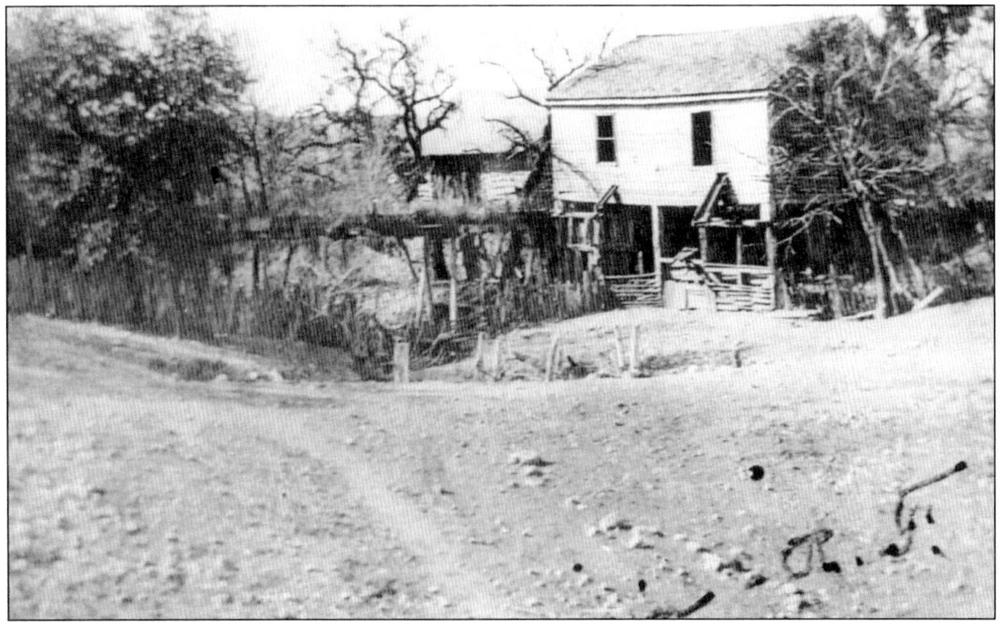

The Rio Frio Mill and Gin was built on the Lombardy Irrigation Ditch by N.M.C. Patterson around 1880. The Lombardy Irrigation Ditch was dug in 1868 with picks and shovels to channel water from the Frio River to irrigate the field crops, gardens, and orchards around the village of Rio Frio. In later years, the ditch became the source of water for mills of the area. (Real County Historical Museum, Leaky.)

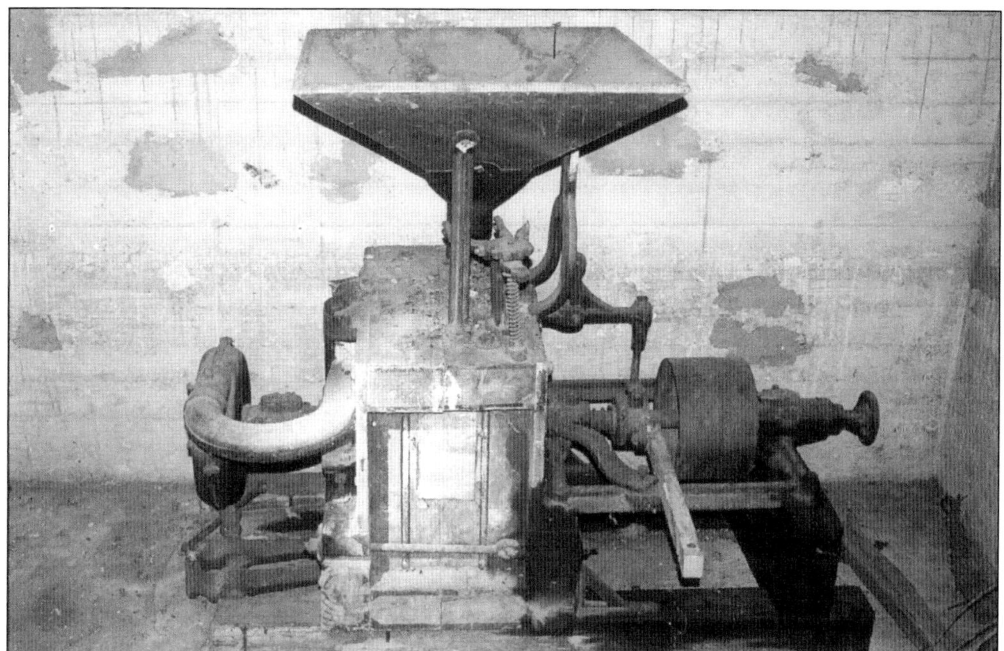

The milling machinery used in the Rio Frio Mill and Gin now sits in a building on the property of Carl Chisum. When the gin went out of business, the mill still ground corn and wheat. The mill had a reputation for grinding the finest flour west of the Pioneer Flour Mill in San Antonio because of its silk flour bolts, which sifted the flour. (Author's collection.)

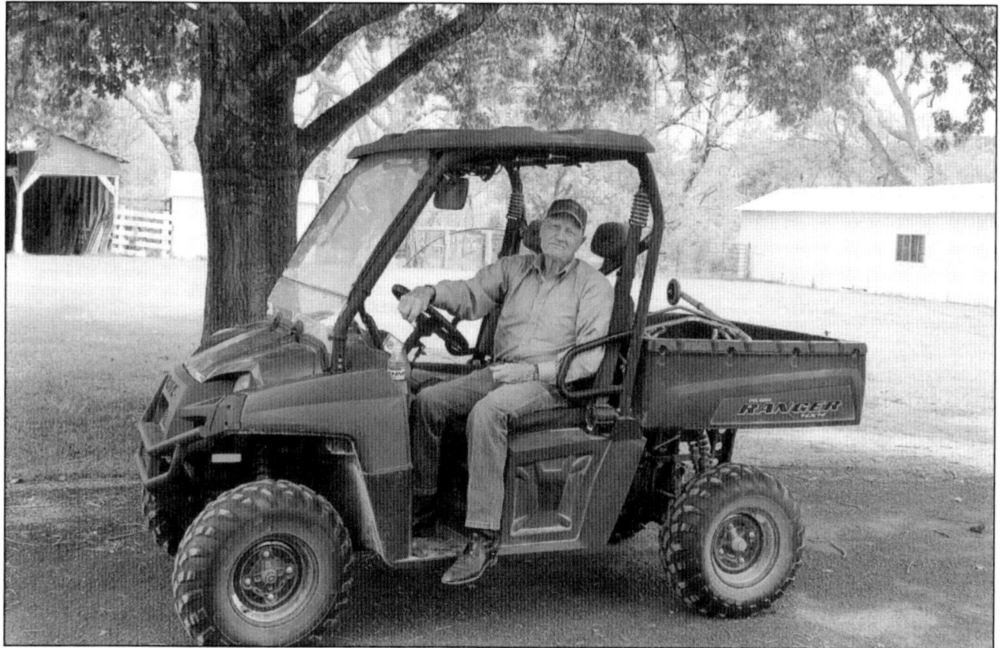

In April 2016, ninety-year-old Carl Chisum gave the author and her husband a tour of the former mill site and pointed out where the two-mile-long Lombardy Irrigation Ditch had run across the back of his property. He also pointed out the route the water took as it entered and left the mill. Chisum has owned the property since 1962. (Author's collection.)

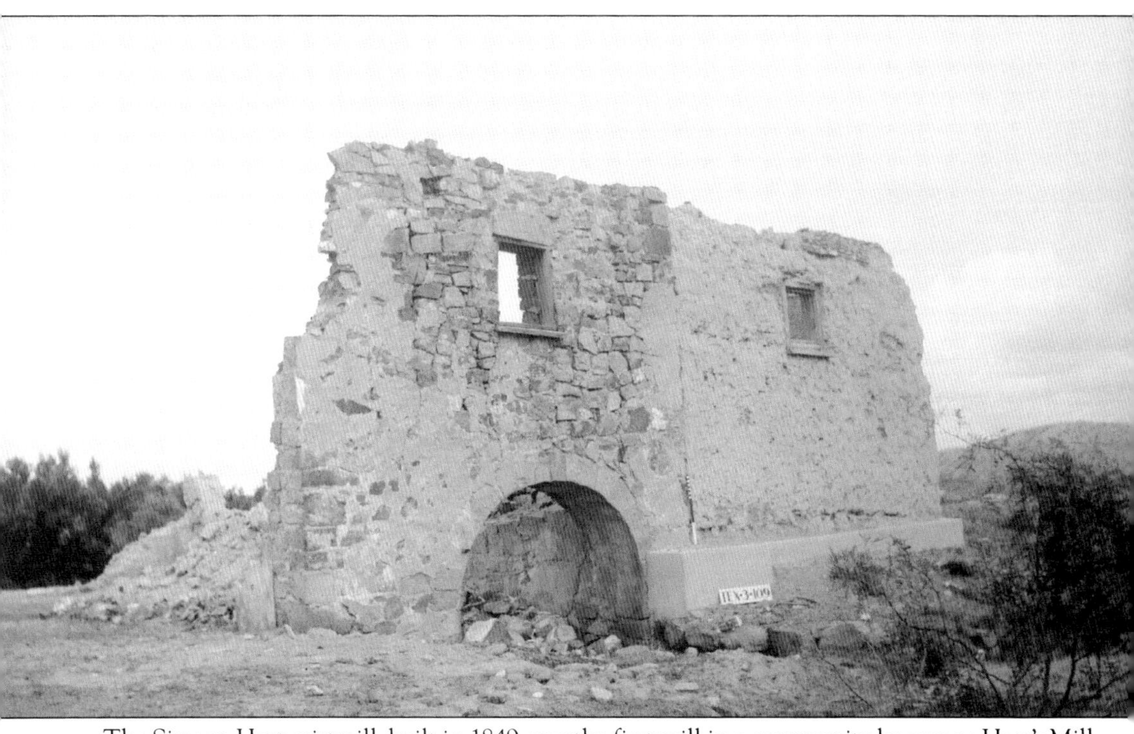

The Simeon Hart gristmill, built in 1849, was the first mill in a community known as Hart's Mill or El Molino. This community grew into a city now known as El Paso in El Paso County. This was the first water-powered mill in the area, with water impounded from the Rio Grande. Power was furnished by a waterwheel, and stones were used for grinding. Grain was brought from hundreds of miles away. The mill had the capacity to produce 100 barrels of flour daily. This flour was shipped as far east as San Antonio, south to Rosales, Mexico, and west to Tucson, Arizona. After Colonel Hart's death in 1874, his son Juan Hart took charge of the mill. In 1880, Juan donated a site adjoining the mill for a military reservation, which became known as Fort Bliss. Hart's Mill supplied the post with flour. The mill continued in operation until 1895 and is considered vital to the early growth of the El Paso area. (Library of Congress, Prints & Photographs Division, HABS TEX, 71-ELPA,1-1.)

Two

The Guadalupe River

Power to a Variety of Mills

The Guadalupe River rises in north and south forks in western Kerr County. After the two branches unite west of Kerrville, the newly formed river meanders through the limestone hills of Kerr County. The river continues to flow southeast for about 230 miles before reaching the Gulf of Mexico. Its principal tributaries are the Comal and San Marcos Rivers.

In the 1700s, several Spanish missions were established along the lower Guadalupe, but they did not survive. The first attempt was in 1726, when Franciscan friars moved an established mission from Matagorda Bay to present-day Victoria, but this effort was abandoned 10 years later. Spanish missions were also established at San Marcos and Comal Springs in 1755 and 1756, but these two missions also failed.

The first successful permanent European settlement occurred in 1824, when Martin de Leon received a grant to colonize lands in the lower Guadalupe valley and established the town of Victoria. Other settlements soon followed. In 1825, four hundred families founded the town of Gonzales. By 1850, Seguin and New Braunfels were becoming important trade centers. The scenic beauty of the river valley and the river itself attracted settlers and many millwrights.

Today the Guadalupe River is used for water sports, but in the 1800s the river was essential for sustaining life. The mills presented in this chapter began at Hunt and followed the river to Kerrville and on to Comfort. Other mills were established near Wimberley, San Marcos, Kendalia, New Braunfels, and Ottine.

Three existing mill sites along the Guadalupe River have undergone a complete restoration. The Gristmill at Gruene (pronounced "Green") is now a popular restaurant, and another mill-turned-restaurant is at the Saffold Dam located at Seguin. The Zedler Mill in Luling is enjoying a new look while keeping its authenticity as a mill complex.

The Guadalupe River, which never seemed to run dry, powered a variety of mills along its banks for several years. However, not all mills were powered by water. The Dutch Windmill at Victoria was powered by wind, and the cottonseed oil mill at Flatonia was powered by steam.

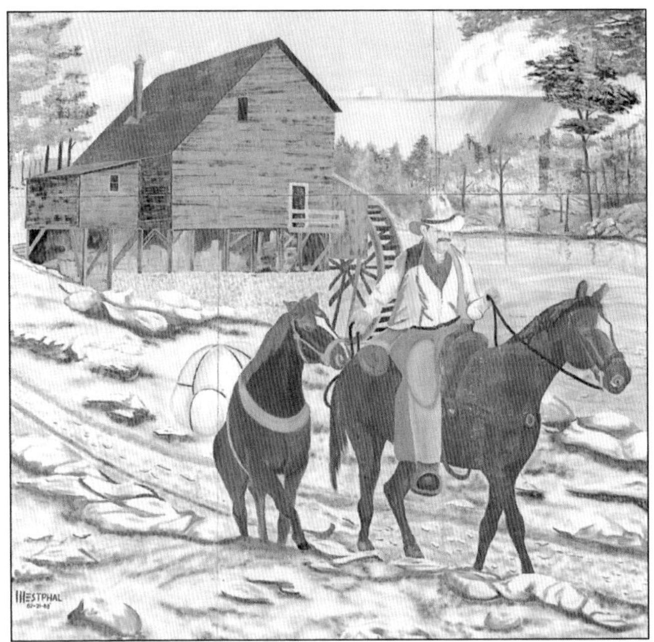

This mural by Jack Feagan pays tribute to the 10 mills that once lined the banks of the Guadalupe River between Hunt and Comfort. The mural is one of 16 panels depicting the history of Kerr County. The panels are mounted on the walls of T.J. Moore Lumberyard in Ingram. Feagan painted the murals in 1989, and restorations were done in 1995 and 2004. (Author's collection.)

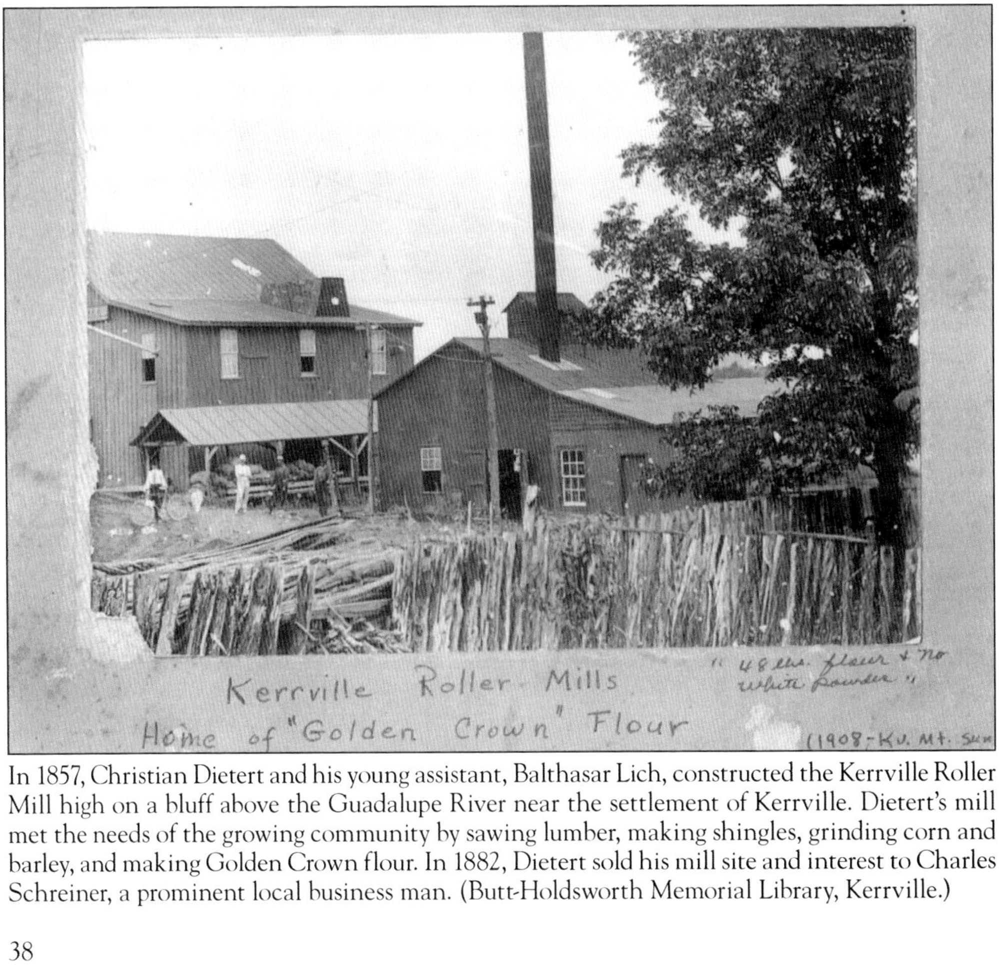

Kerrville Roller Mills
Home of "Golden Crown" Flour

In 1857, Christian Dietert and his young assistant, Balthasar Lich, constructed the Kerrville Roller Mill high on a bluff above the Guadalupe River near the settlement of Kerrville. Dietert's mill met the needs of the growing community by sawing lumber, making shingles, grinding corn and barley, and making Golden Crown flour. In 1882, Dietert sold his mill site and interest to Charles Schreiner, a prominent local business man. (Butt-Holdsworth Memorial Library, Kerrville.)

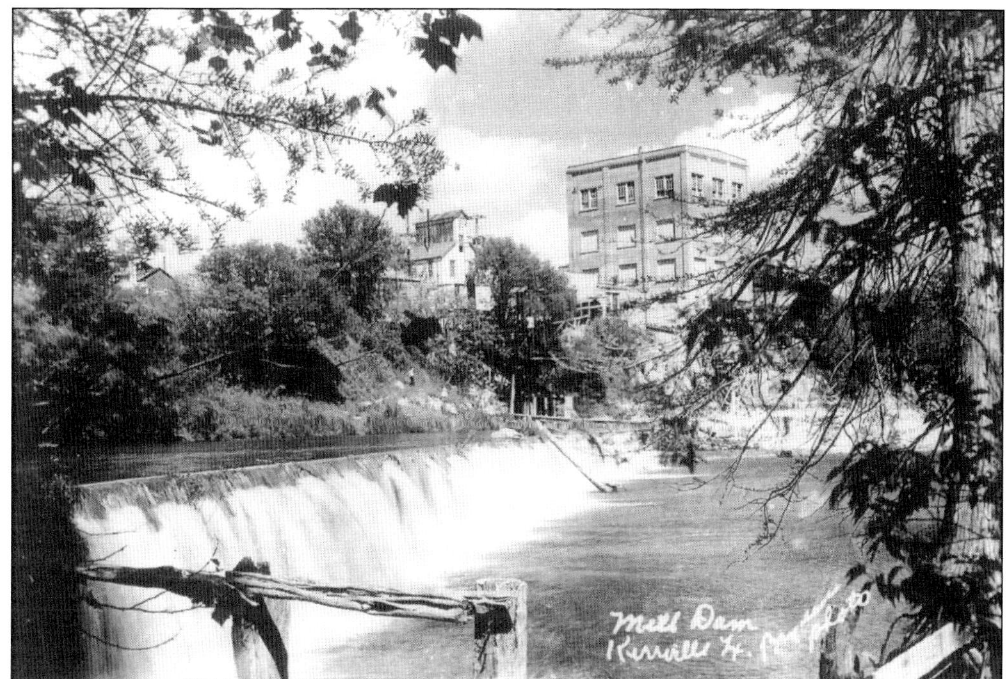

When Charles Schreiner purchased the Dietert Mill, he continued to operate the flour mill, shown at left. He also installed equipment for operating an electric power plant, the large building to the right. An underwater iron turbine, which Dietert had installed in 1868, powered the mill and the power plant. The Schreiner One Center office complex now occupies the site of the former power plant. (Butt-Holdsworth Memorial Library, Kerrville.)

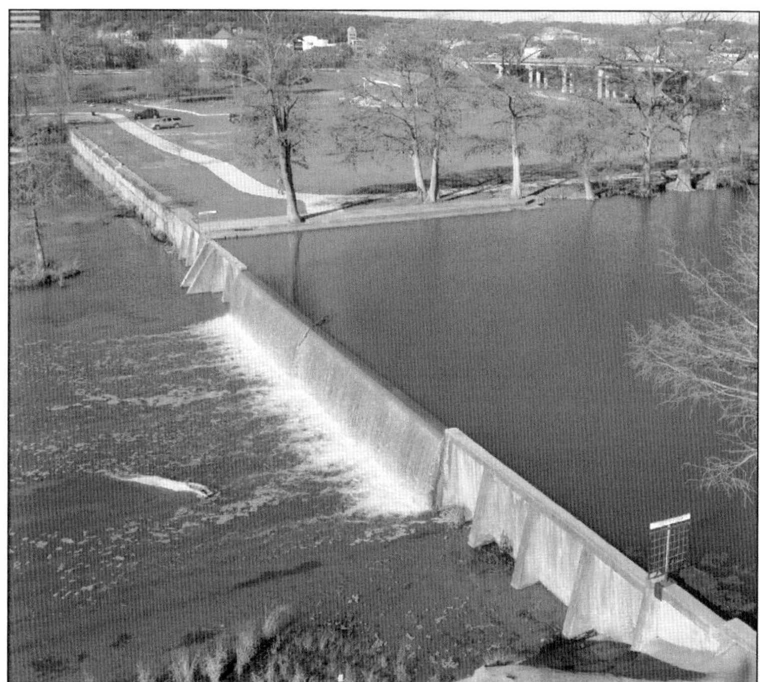
This photograph of the present-day Guadalupe River and dam was taken from the site now occupied by the Schreiner One Center office complex. In its early days, the river fostered the development and economic growth of the city of Kerrville and other historic villages along its banks. Today, the Guadalupe River is considered by many to be the best recreational river in Texas. (Author's collection.)

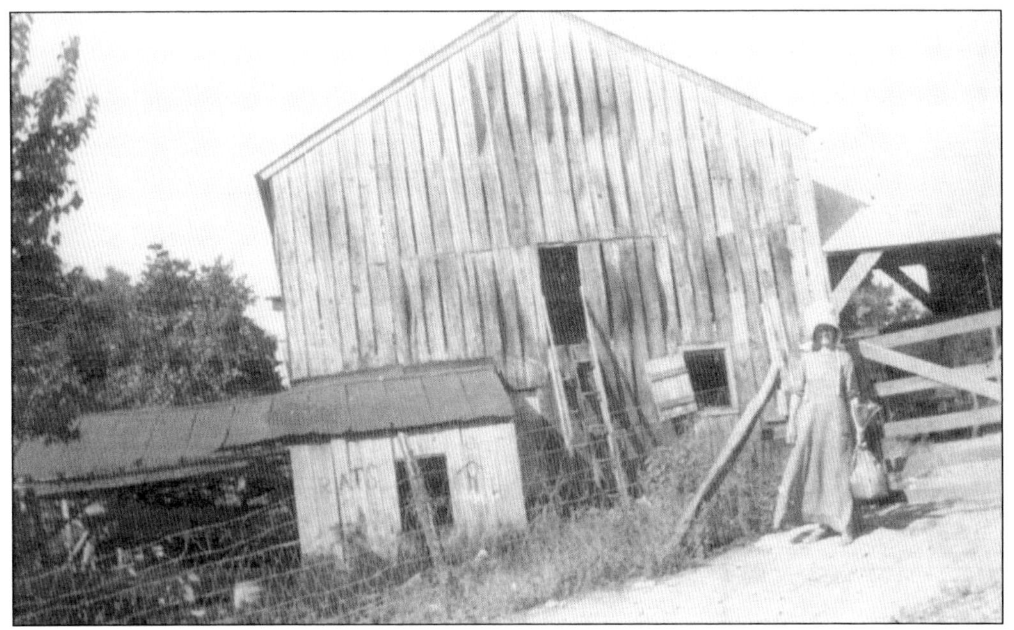

Wimberley Mill was built in 1848 by William C. Winters as a gristmill and sawmill on Cypress Creek, Hays County. After a flood destroyed the millhouse in 1856, Winters moved to higher ground across the creek and built a two-story millhouse with a long millrace and tailrace. Upon Winters's death in 1864, the mill passed to his son-in-law, John M. Cude. (San Marcos–Hays County Collection, San Marcos Public Library.)

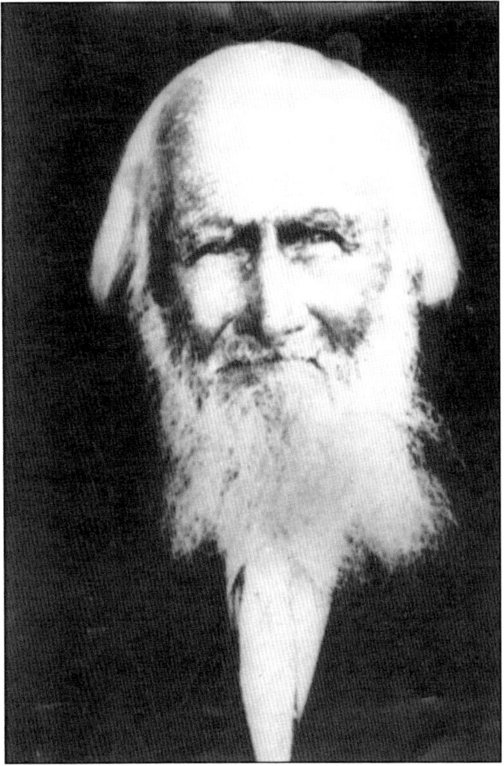

In 1874, Pleasant Wimberley bought the mill tract, which included the mill and Winters's home. He partnered with his son Zachariah and grandson Calvin, and the milling operation grew to include a gristmill, sawmill, shingle mill, molasses mill, and one-stand cotton gin. This thriving business soon became the economic heart of the village that became known as Wimberley Mills, now known as Wimberley. (San Marcos–Hays County Collection, San Marcos Public Library.)

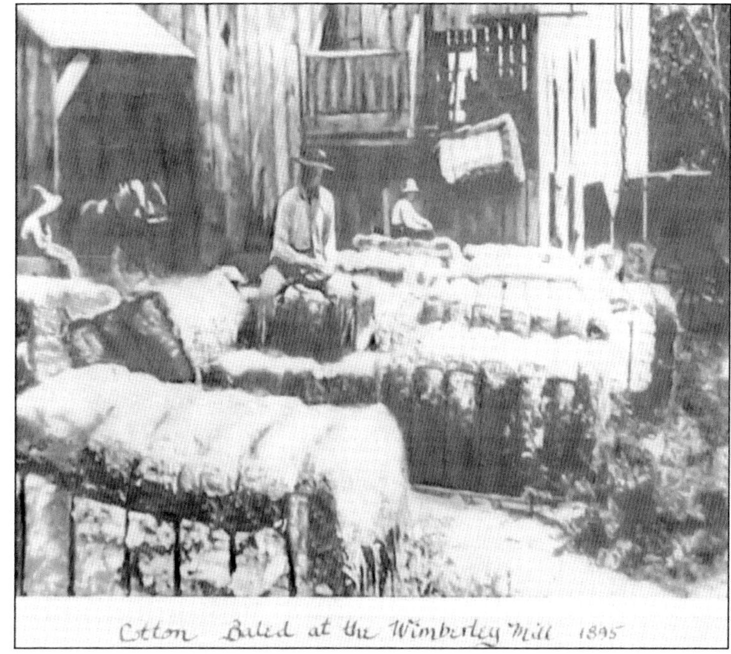

This photograph shows cotton that was baled at the Wimberley Mill in 1895. Normally, 300 bales were ginned in a season, but in 1900, the rains came at the right time, and that season, over 900 bales were ginned, with a profit of $1 per bale. The charge for baling was $3 per bale. (Wimberley Institute of Cultures Historic Photograph Collection.)

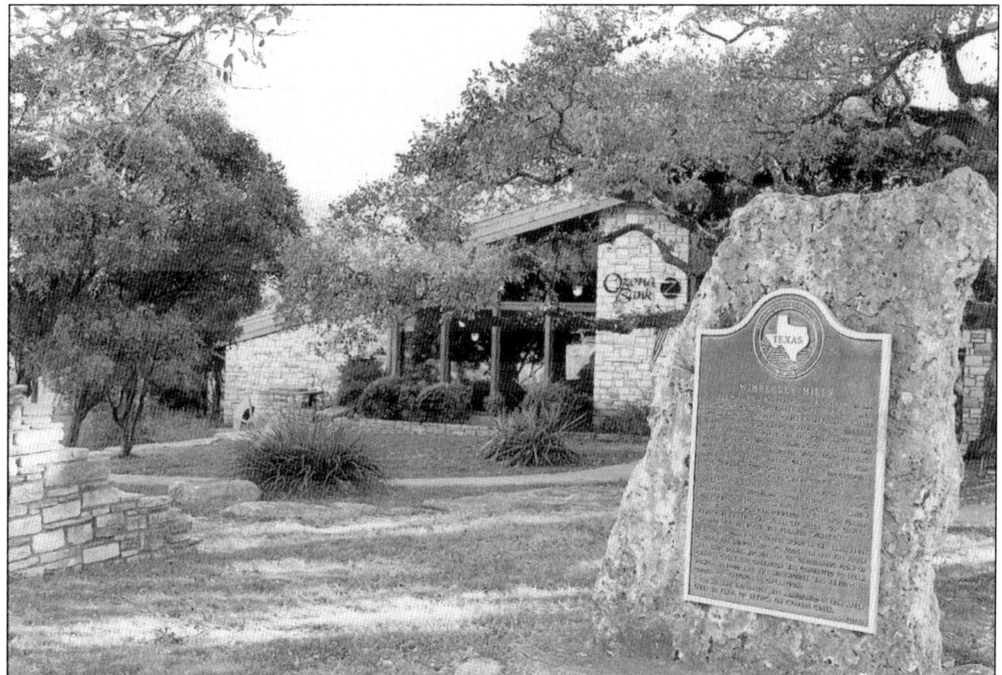

Pleasant Wimberley died in January 1919. Inscribed on his tombstone are the words "Pleasant Wimberley—For Whom the Town Was Named." Other family members kept the mill going for several years following Wimberley's death, but the days of the small-town custom mill soon drew to a close. The mill ceased operation in 1925 and was razed in 1934. A historical marker and the Ozona Bank now sit on the site of the Wimberley Mill. (Author's collection.)

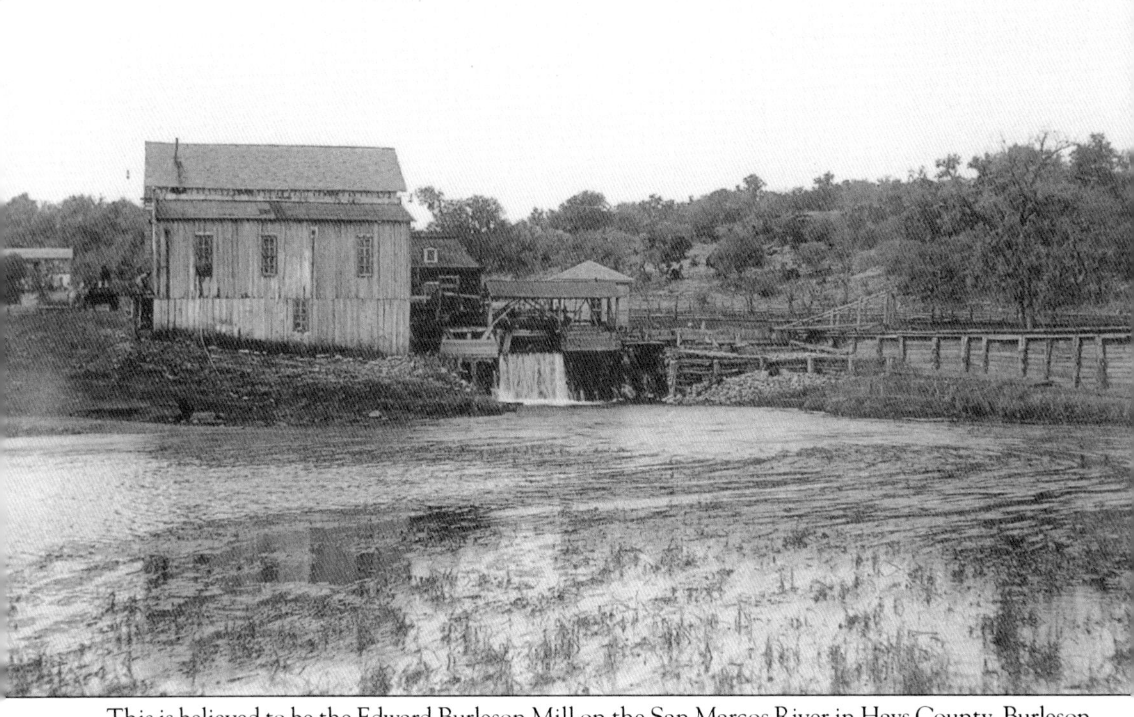

This is believed to be the Edward Burleson Mill on the San Marcos River in Hays County. Burleson, a native of North Carolina, moved to Texas in 1830, settling first in Bastrop County. Burleson moved his family to the San Marcos River valley in the late 1840s. It was here that he built the first dam on the river to provide power for a gristmill and sawmill, which became known in later land transactions as the "Mill Tract." This Mill Tract evolved into the center of a commercial enterprise for the City of San Marcos. In years to come, this tract of land would house a cotton gin, three ice factories, a water works, and an electric plant. This early industrial park, while no longer in existence, had a steady economic impact on the town of San Marcos for over a century. (MSS0717-0237, Houston Public Library, Houston Metropolitan Research Center.)

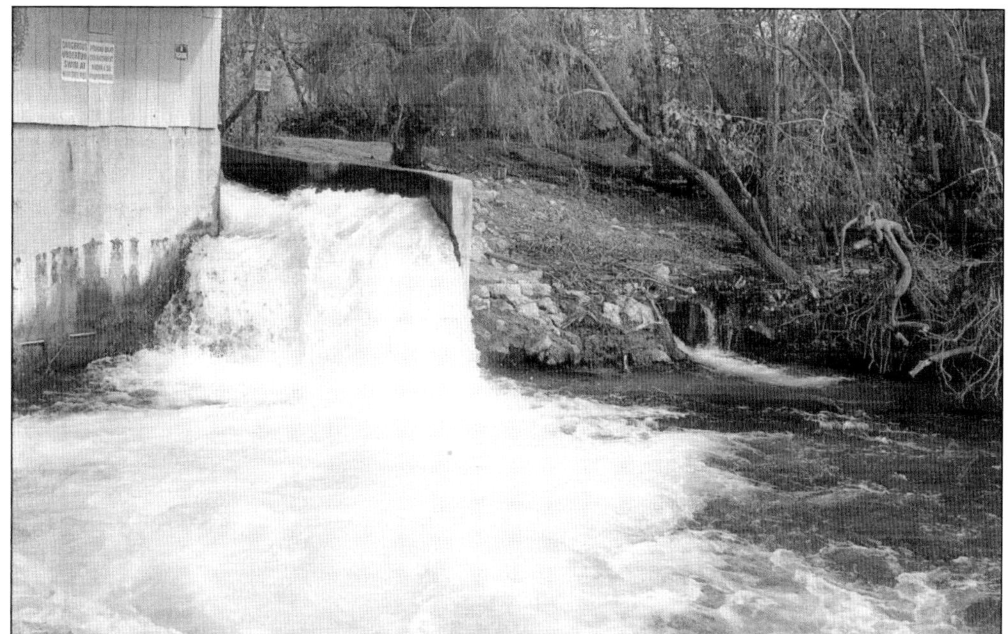

A popular restaurant now occupies the site of the former Edward Burleson Mill. A Texas Historical Commission marker standing in front of the restaurant commemorates the significance of the Burleson Mill and the economic impact it had on the growth of San Marcos's early Industrial Park and the town of San Marcos. (Author's collection.)

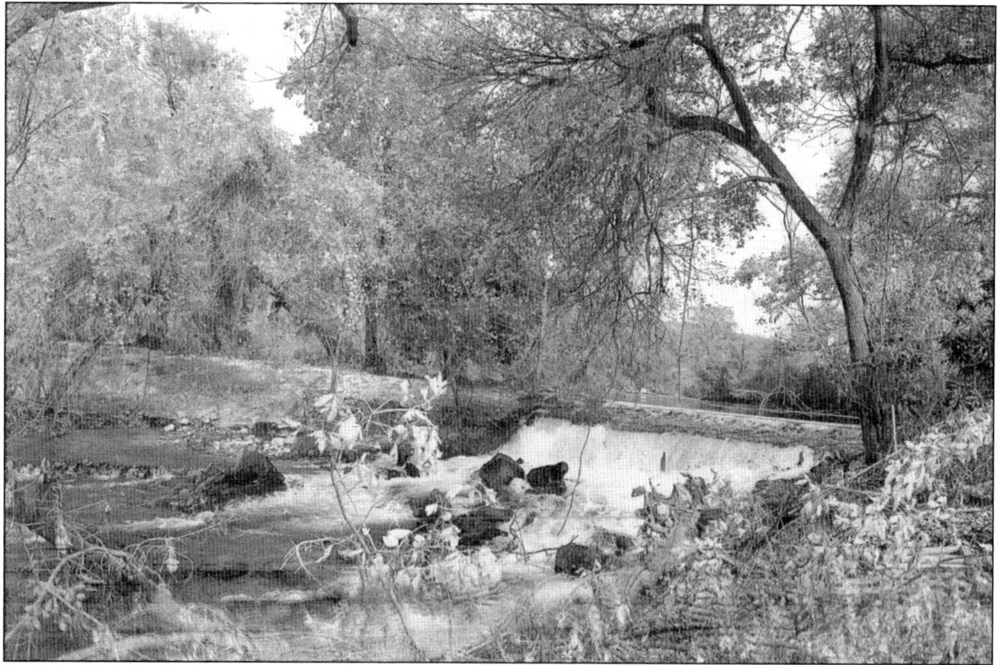

This scenic dam on the San Marcos River is very near the original dam erected to provide power for the Burleson Mill. The Burleson Mill is the oldest of five mill sites within the city of San Marcos. Others were the Samuel L. Pegues, W.W. Woolfork, Charles L. McGehee Jr., and Tom Code. (Author's collection.)

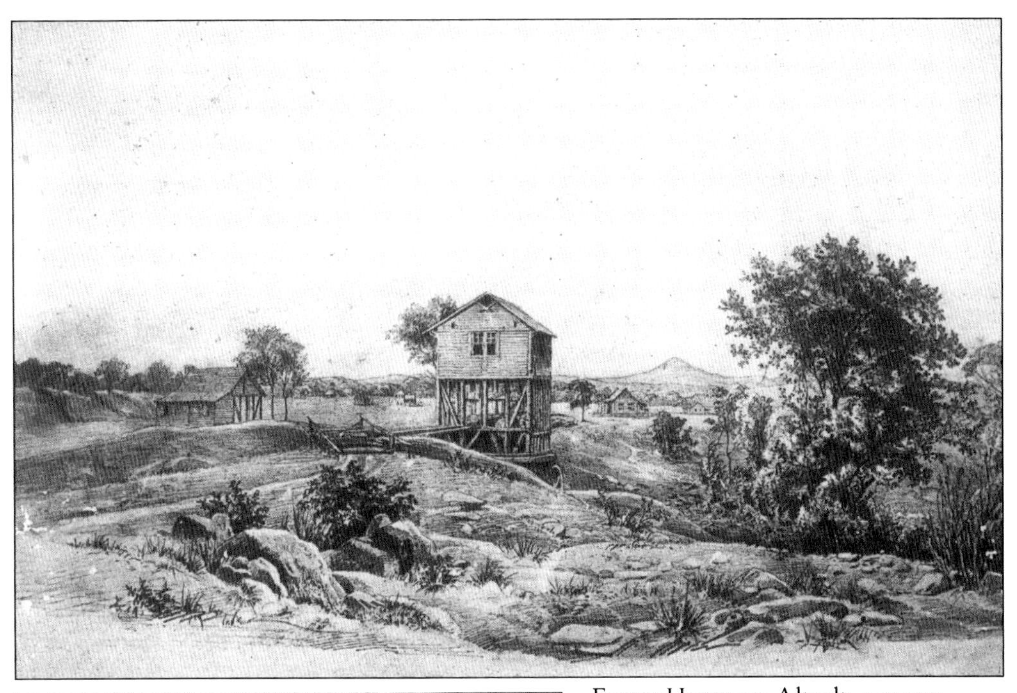

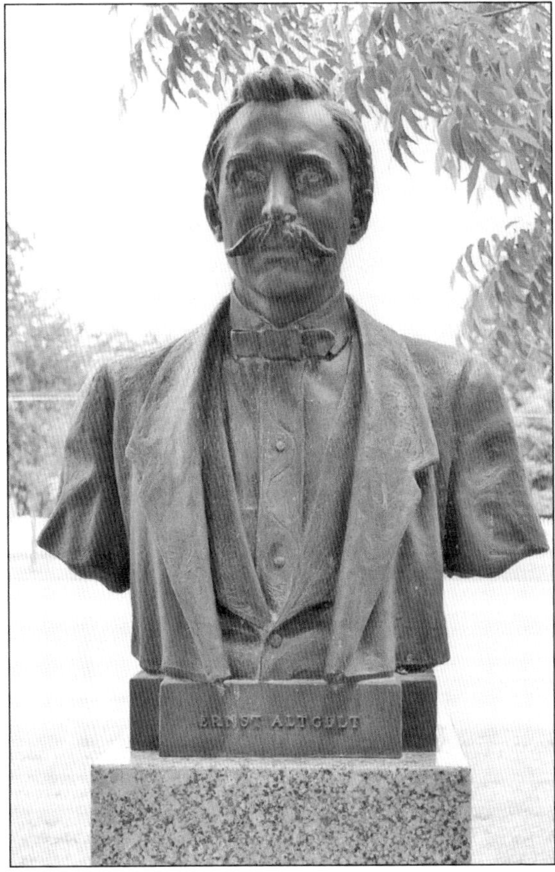

Ernest Hermann Altgelt, a young German immigrant to New Orleans, worked in the cotton firm of John Viles. In 1854, Altgelt led a surveying party into Texas and laid out the town of Comfort, Kendall County, on property owned by Viles. Altgelt then commissioned Christian Dietert to build a lumber and gristmill. This drawing of Altgelt's Perseverance Mill is by Hermann Lungkwitz. (The University of Texas at San Antonio Libraries, Special Collections.)

This bronze statue of Ernest Hermann Altgelt (1832–1878) was placed in the Comfort City Park in recognition of Altgelt's contributions in founding the town of Comfort. Altgelt proclaimed the founding date as September 3, 1854. He erected his sawmill and gristmill to assure the economic stability of the new settlement. Altgelt also donated the land on which the park was later built and developed. (Author's collection.)

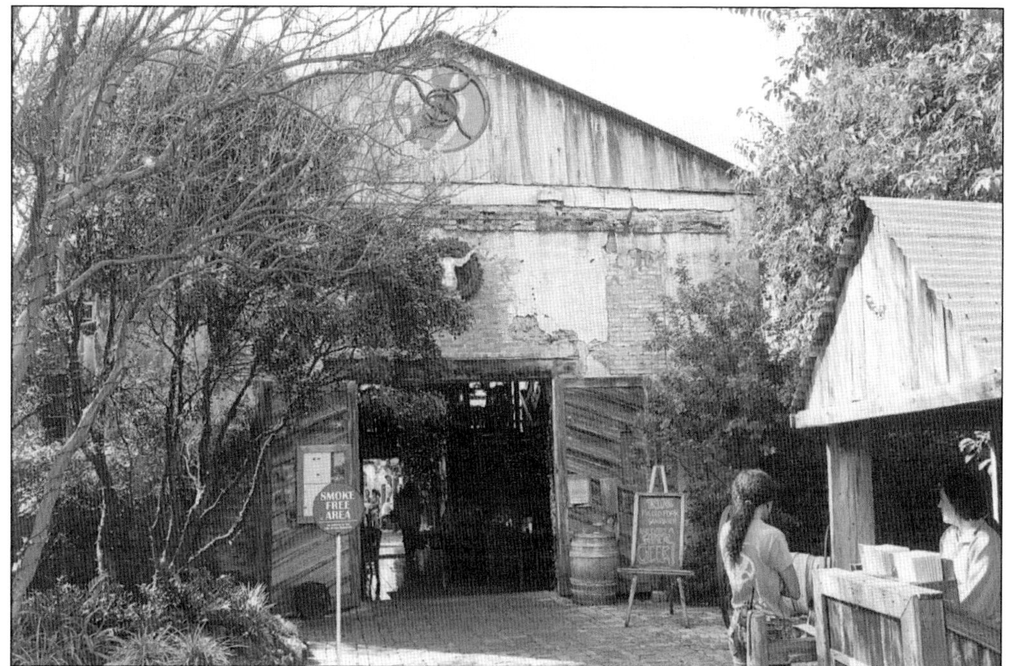

In 1872, Henry D. Gruene of New Braunfels, Comal County, founded the town of Gruene by establishing a waterwheel gristmill powered by the Guadalupe River. He also constructed a steam-operated gin to process the vast amounts of cotton being grown in the area. The cotton gin burned in 1922, with only a shell of itself remaining. After repairs were made, the building became a popular destination restaurant called the Gristmill. The millstones shown below are from the Henry D. Gruene gristmill, the first business established in the township of Gruene. Gruene was a land developer, farmer, and rancher, and he owned land from the Red River to the Gulf of Mexico. This marker sits in front of the Gruene Family Home and was dedicated in 1984 to Henry D. Gruene and Bertha Gruene by their grandchildren. (Both, author's collection.)

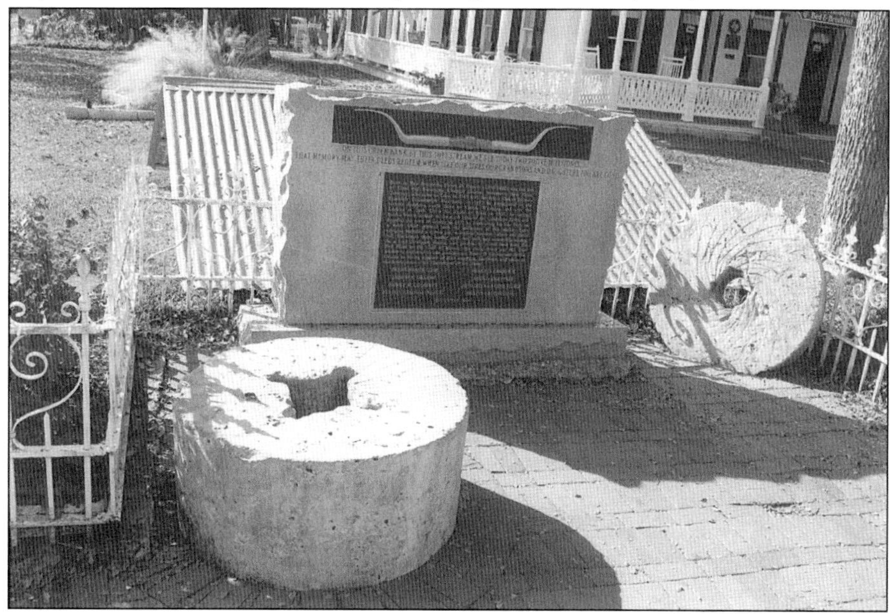

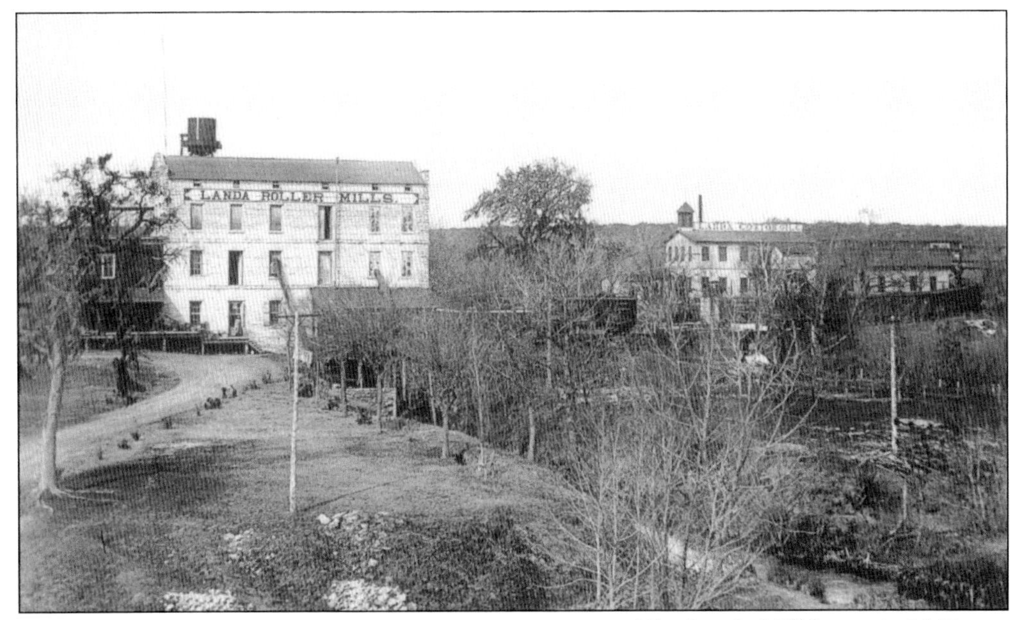

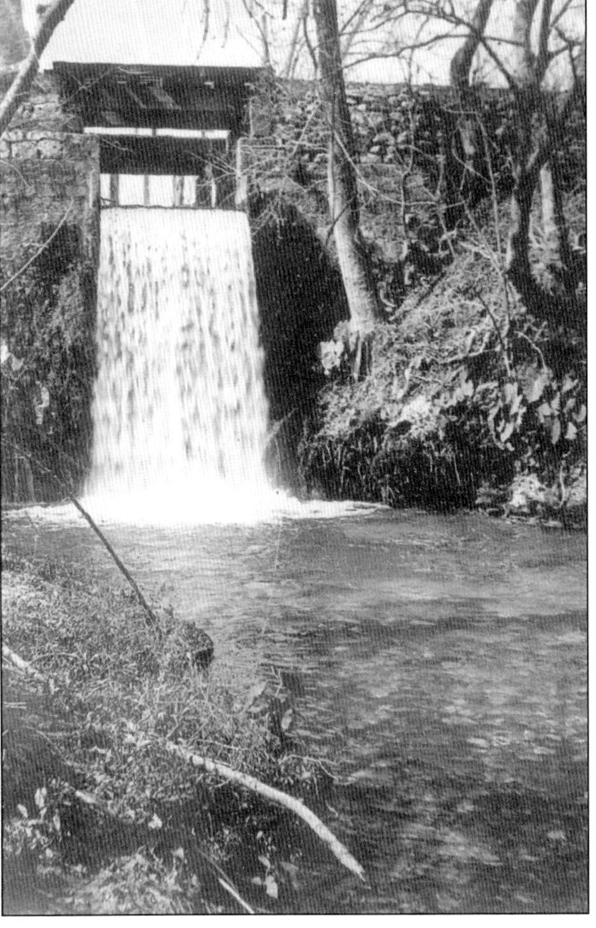

The Landa Mill began in 1847, when William Meriwether purchased land in New Braunfels, Comal County, where he built a gristmill, cotton gin, and sawmill. In 1859, Meriwether sold his holdings to Joseph Landa. Landa and son Harry expanded the milling operation to include a flour mill, cottonseed oil complex, ice plant, and hydroelectric plant. Landa Industries was a major contributor to the rapid growth of New Braunfels. (MSSO717-0246, Houston Public Library, HMRC.)

Water flows over the dam at the Landa Milling Company Plant. Landa Industries continued to operate until the family began to liquidate its assets in the mid-1920s. In 1936, after being purchased by the city, much of the original Meriwether property became a popular recreation area known as Landa Park. (MSS0717-0241, Houston Public Library, HMRC.)

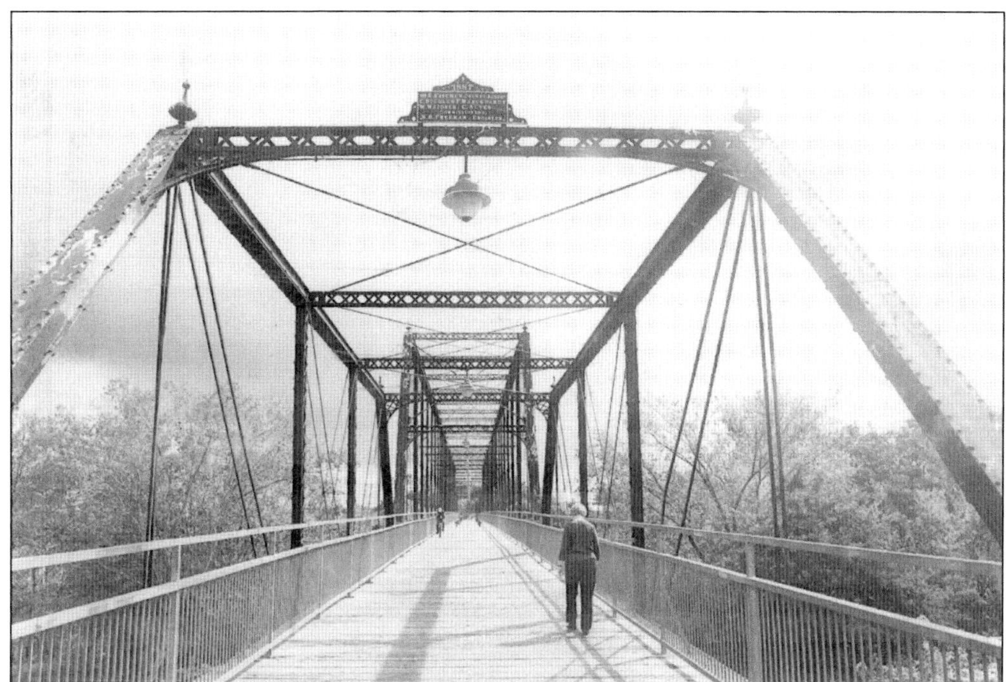

The Faust Street Bridge in New Braunfels, constructed in 1887, opened up the east bank of the Guadalupe River for settlement and development. Prior to the bridge, caravans carrying supplies to Spanish missions in East Texas and travelers crossing the river on El Camino Real had to ford the river. The bridge made the ford unnecessary, and a mill dam, constructed in 1923, completely submerged the ford. One of the first developments on the east bank of the river was a textile mill originally named the Planters and Merchants Mill. The photograph below shows the dam, the millhouse that was built to house the generators, and the mill. Later, the mill was known as the Mission Valley Mills, and it became a major producer of Bluebonnet gingham and other cotton fabrics. The mill opened in 1923 and closed in 2004. (Both, David Carson.)

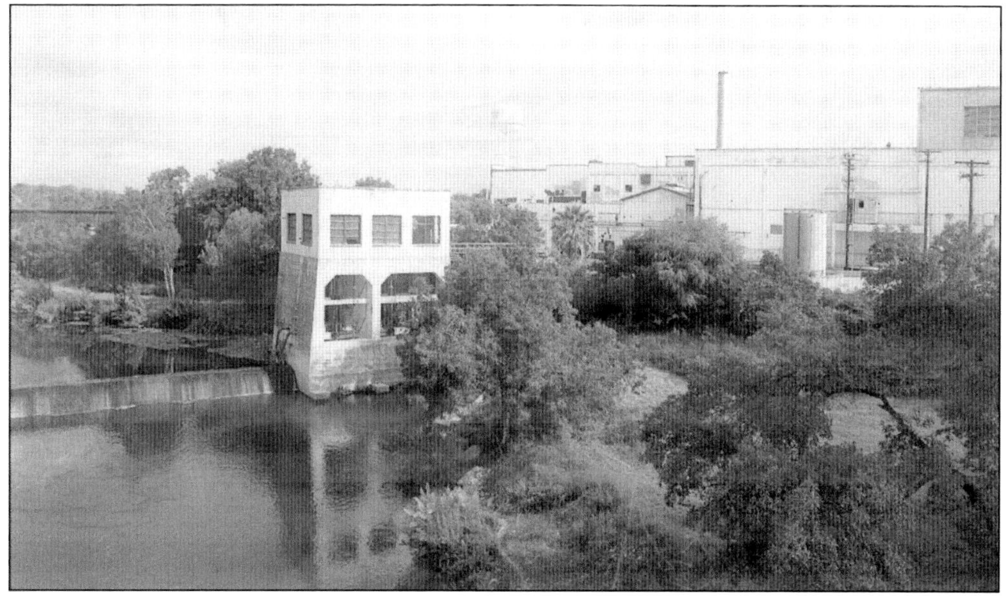

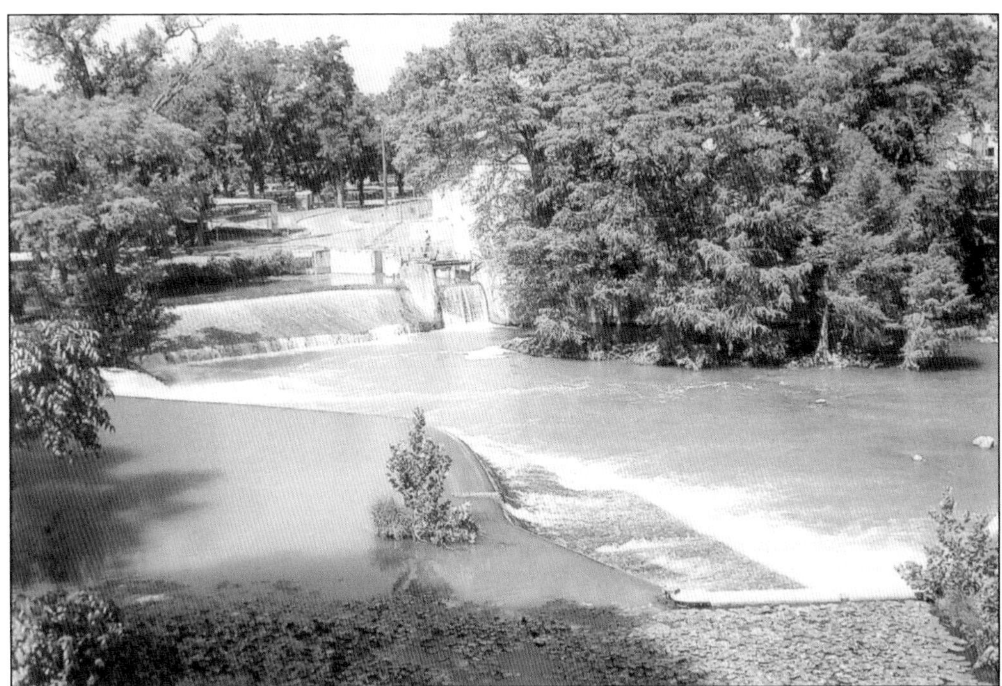

Saffold Dam, on the Guadalupe River near Seguin, Guadalupe County, was named for Civil War veteran William Saffold, who owned much land in the Seguin area. The dam originated from a natural S-shaped outcropping that Saffold improved by adding boulders. In 1879, Henry Troell purchased land near the river, including the Saffold gristmill. A portion of the old millrace is visible in the background above. (Author's collection.)

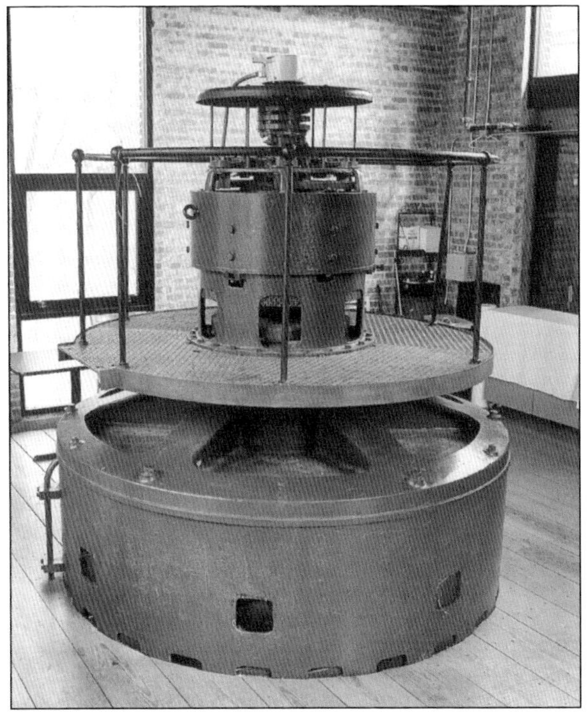

In 1866, Troell installed a 54-inch turbine wheel at his power dam. After the installation of the turbine, he began pumping water to supply the city of Seguin and vicinity. The Troell power plant ceased operation years ago and is now a destination restaurant called the Power Plant Texas Grill. Guests may dine in the old turbine room, which has been fully renovated for private parties. (Author's collection.)

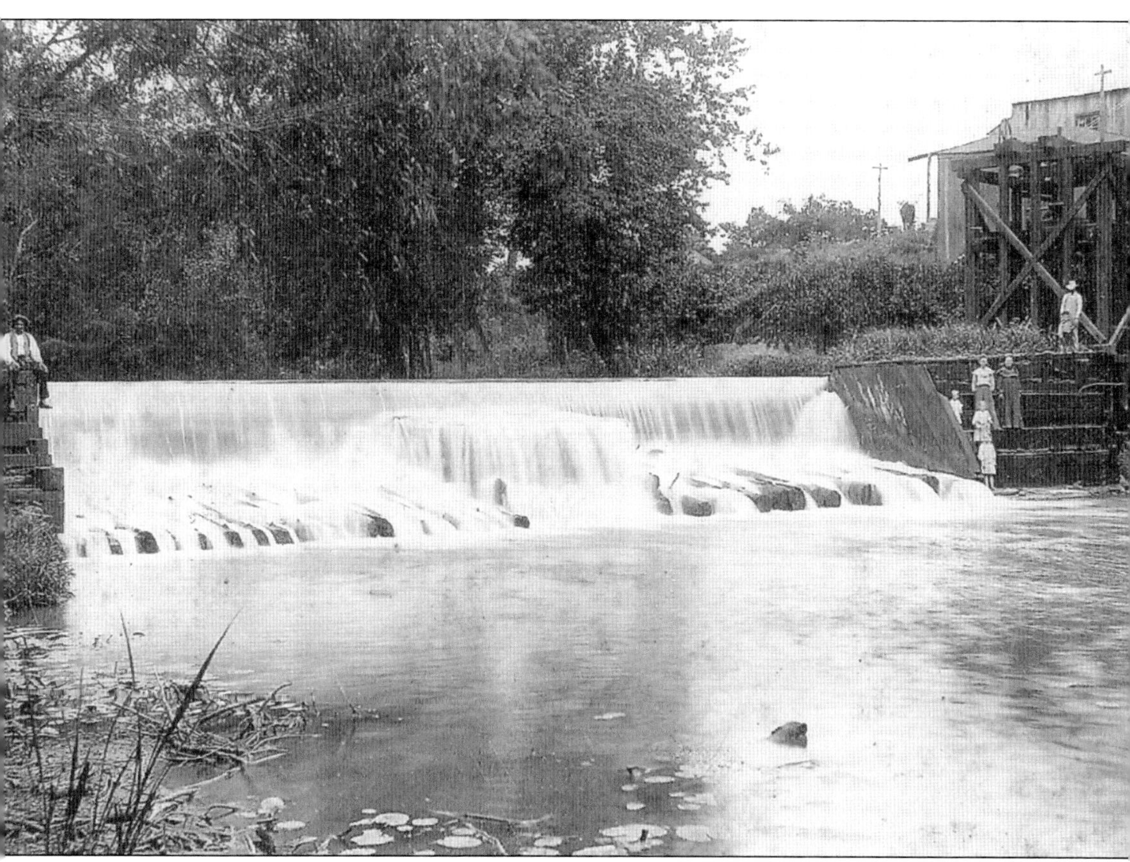

The San Marcos River near Luling, Caldwell County, provided power for the Meriwether gristmill, which was built near its banks in 1874, the same year the town of Luling was established. The mill and a simple stone dam across the river were built by Leonidas Hardeman and two brothers, John and James Meriwether, all from Tennessee. Soon after the gristmill was established, the three owners added a gin, because the land around Luling yielded an abundance of cotton. They also installed a waterwheel at the dam to produce power for their machine shop. This gave Luling enough water to supply homes and businesses. The mill was known as the Meriwether Mill. Hardeman and the Meriwether brothers operated the mill at a loss for several years before relinquishing their ownership to four enterprising investors. (Zedler Mill Foundation, Luling.)

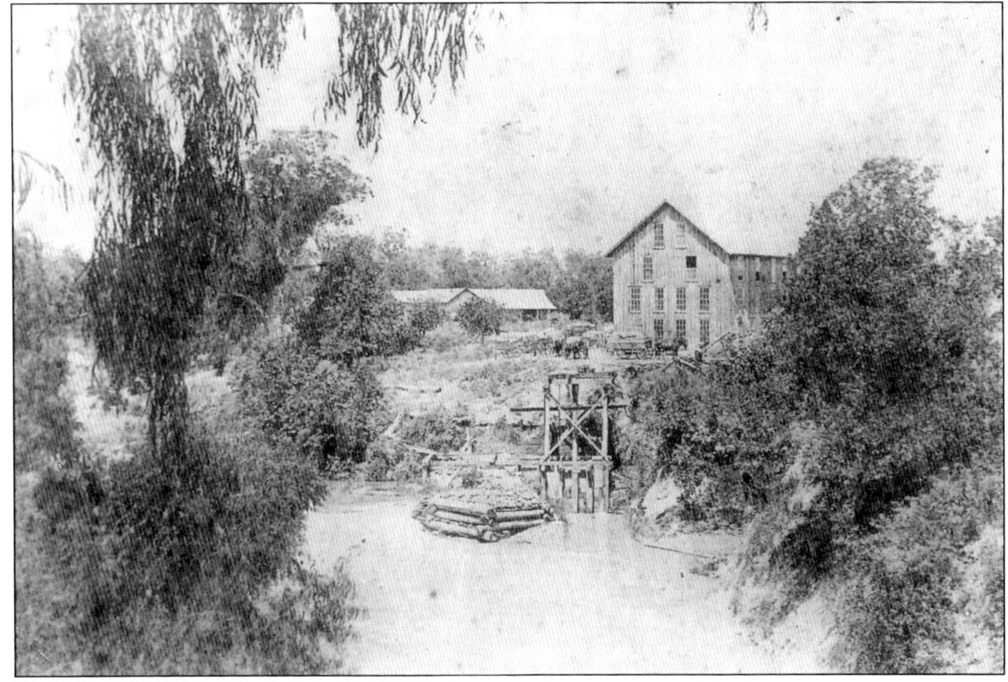

In 1884, Bob Innes, John Orchard, J.K. Walker, and Frederick (Fritz) Zedler, a German immigrant millwright, purchased the site and mill equipment, naming it the Luling Water Power Company. The partners added a lumber sawmill, and Zedler replaced the stone dam with a wooden dam and penstocks that could generate more power. Within a few years, Zedler bought out his three partners and became sole owner of the mill. (Zedler Mill Foundation, Luling.)

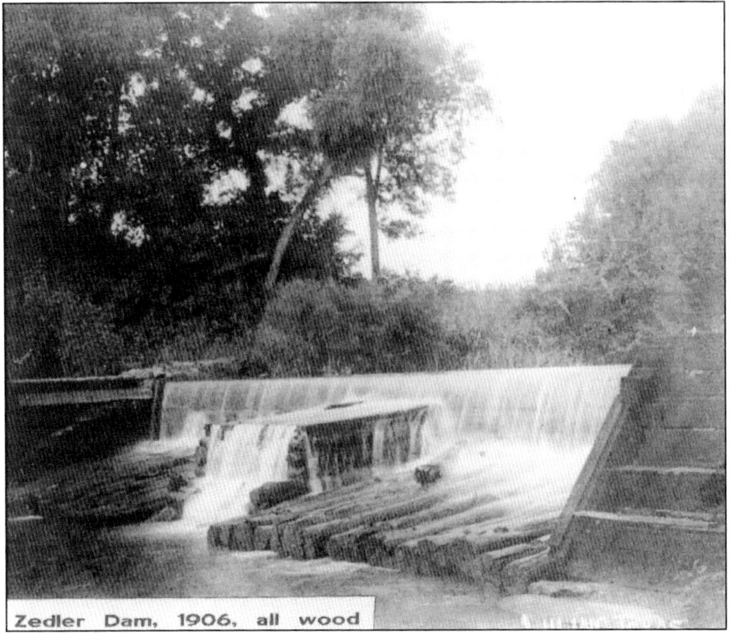

This is a view of the wooden dam that replaced the original stone dam. After Zedler became the sole owner in 1888, his oldest son, Berthold, became his partner. A few months later, on October 15, 1888, the entire three-story mill building burned to the ground. The Zedler family met this challenge head-on, and within seven weeks, a new and improved mill factory was in operation. (Zedler Mill Foundation, Luling.)

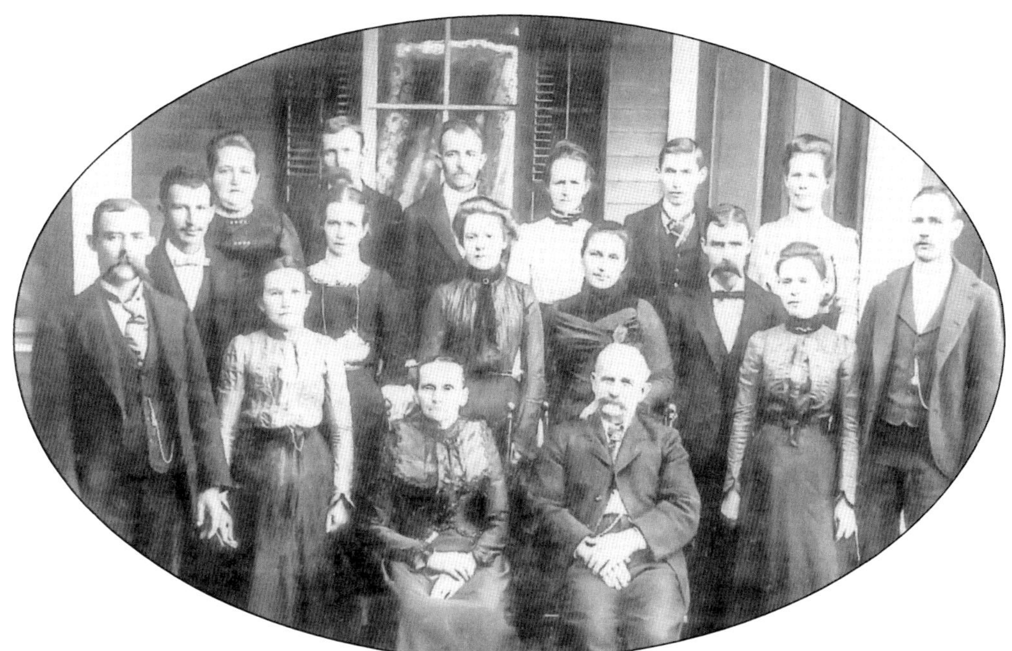

The Fritz Zedler family is, from left to right, (first row) Berthold and Marie Zedler, Louise and Fritz Zedler, and Jane and Caesar Bunde; (second row) Fritz and Helen Bohne, Frances Zedler, Lottie and Charles Zedler; (third row) Emma and Herman Zedler, Charles and Louise Gips, and Carl and Pauline Eckhardt. (Zedler Mill Foundation, Luling.)

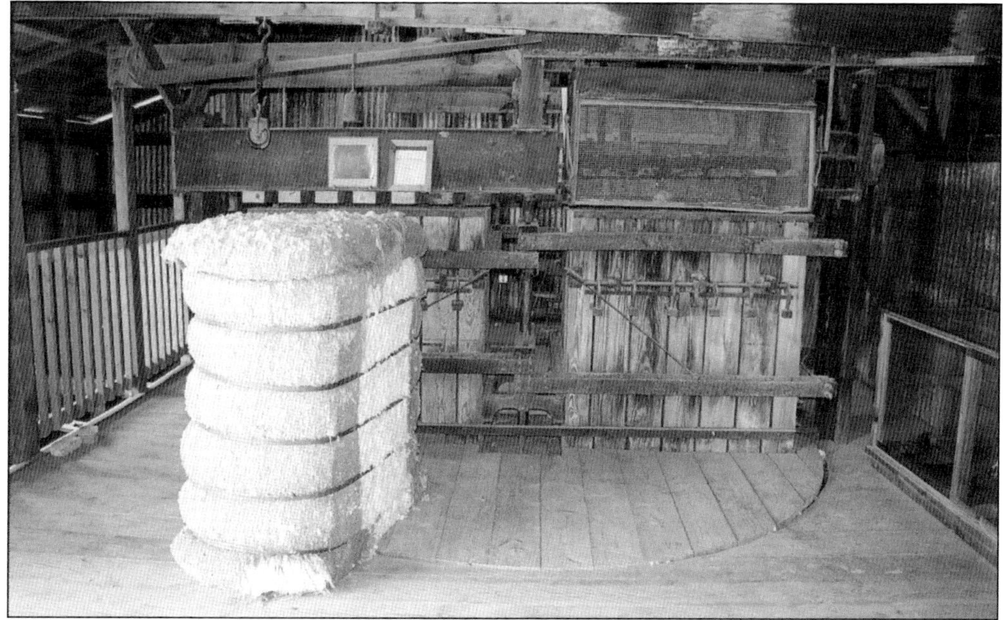

During its years of operation, cotton farmers brought in 1,500-to-1,800-pound wagonloads of cotton along with dirt and seeds. The cotton was sucked up into the mill with vacuums and went through a cleaning process. The cleaned lint was then compressed into 500-pound bales ready to ship to textile mills. This photograph shows a box press, where the lint was squeezed and banded into a bale of cotton. (Author's collection.)

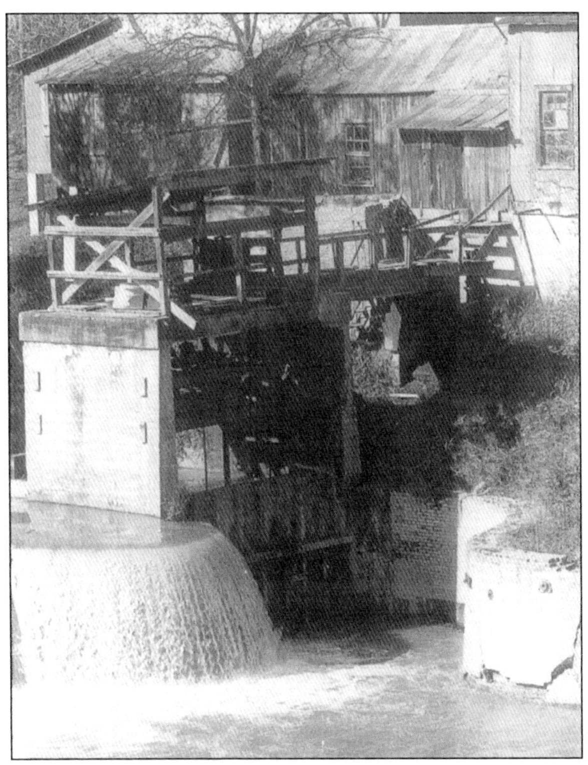

This is a view of the millrace and water turbine at the Zedler Mill. Each generation of Zedler family owners improved the cotton gin and gristmills and lumber mills by adding water turbines, steam engines, a concrete dam, mule barns, and a blacksmith shop. In 1894, sons Herman and Charles Zedler installed a generator to supply the town of Luling with electric power. (Zedler Mill Foundation, Luling.)

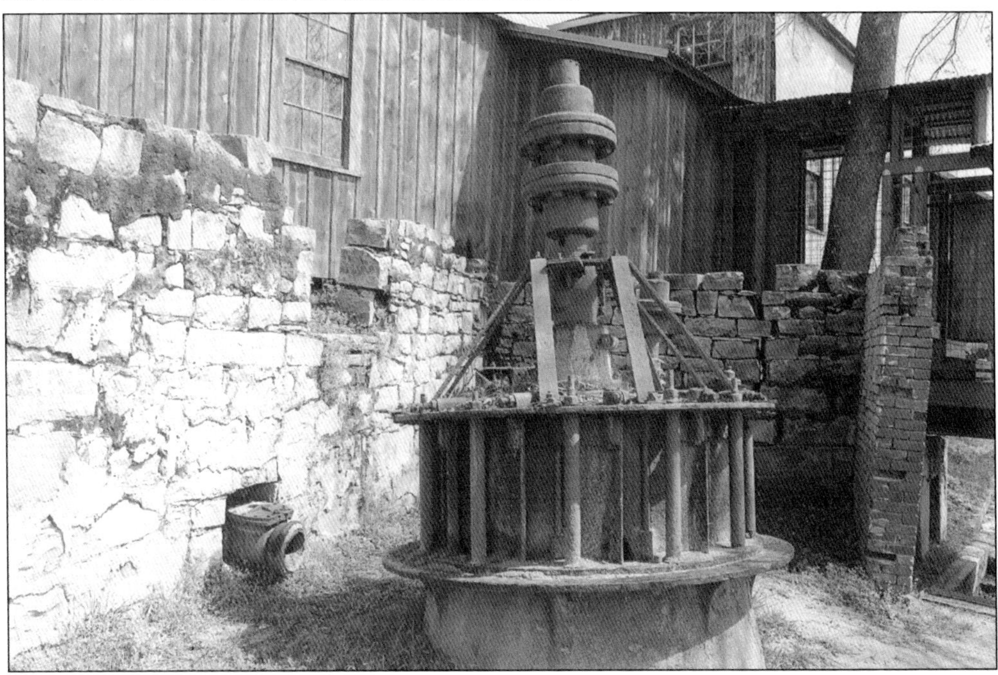

Zedler Mill enjoyed an 80-year history as a cotton gin, gristmill, and lumber mill before sadly relinquishing itself to the modern methods of milling and then falling into a state of disrepair. After operations ceased in the late 1950s, all the mill equipment and power machinery, with the exception of the turbine shown above, was sold for scrap. (Author's collection.)

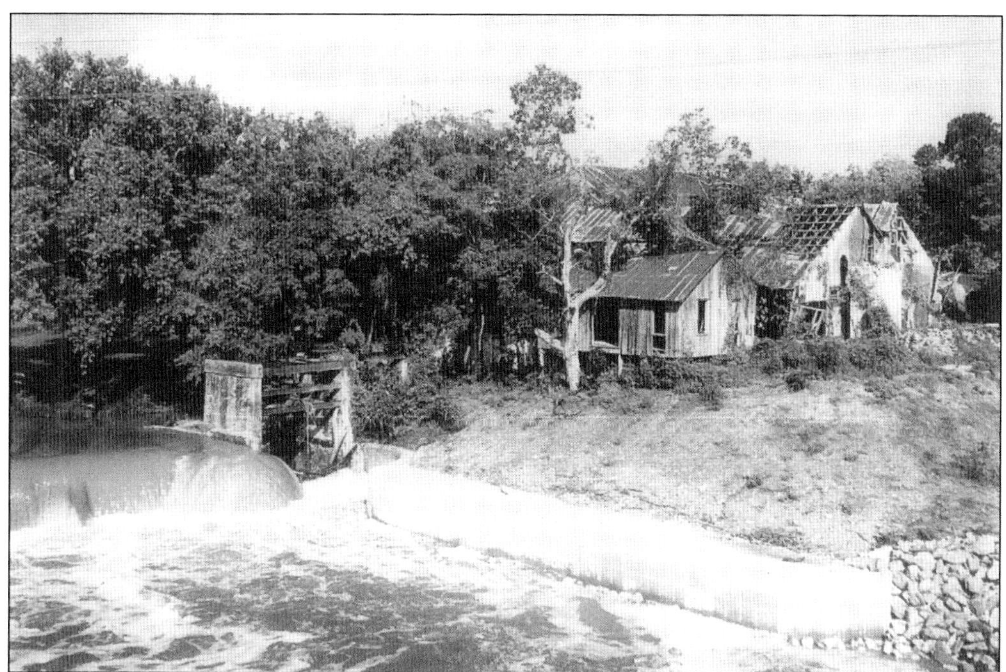

This October 26, 2004, photograph shows the state of disrepair the mill buildings suffered when the mill ceased operating. In 2002, the Luling Economic Development Corporation purchased the property for the City of Luling. The Zedler Mill Foundation was formed in 2007 to restore and develop the mill site. (Zedler Mill Foundation, Luling.)

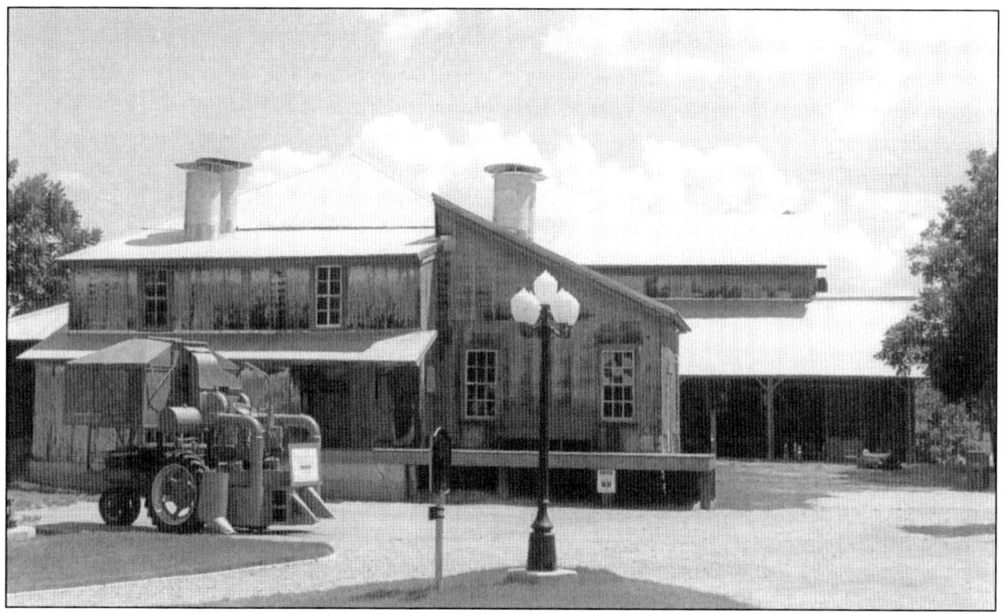

This photograph shows a portion of the newly restored mill and cotton museum, which visitors may tour at their leisure. The restored scale house, once used to weigh wagons full of produce, serves as a hub for the complex. A mule barn, produce storage building, and other outbuildings are all available for viewing. This historical complex is open year-round, and there is no admission fee. Visitors are always welcome. (Zedler Mill Foundation, Luling.)

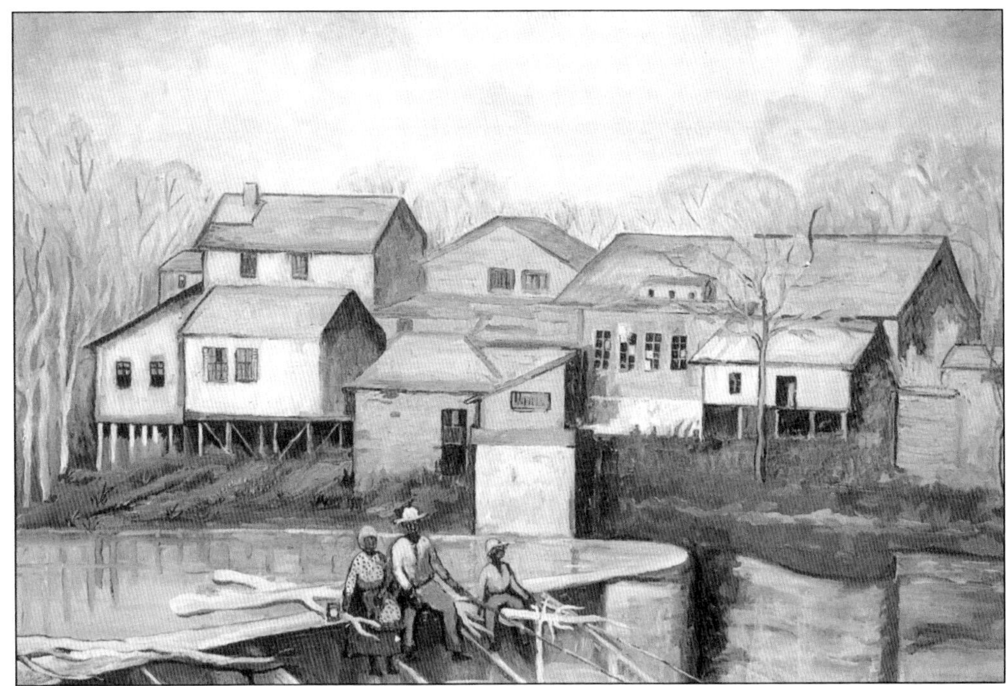

This painting of the Zedler Mill was done by folk artist Ruby Hayes Palmo. The painting shows the numerous additions that were made to the mill as it changed to meet the ever-growing needs of the community. In the early 1950s, a woman passing through Luling saw the painting in a café and was attracted to it, so she purchased it. After the woman's death in 2001, her daughter, Marty Cassady, knowing how much the painting meant to her mother, brought it home with her. The painting now hangs in the daughter's home in Salado. The Fritz Zedler home, shown below, was built in 1900 according to Zedler's own plans. The original structure had 10 rooms and four porches. The home is located across the street from the mill complex and is available for daily or weekly rental. (Both, author's collection.)

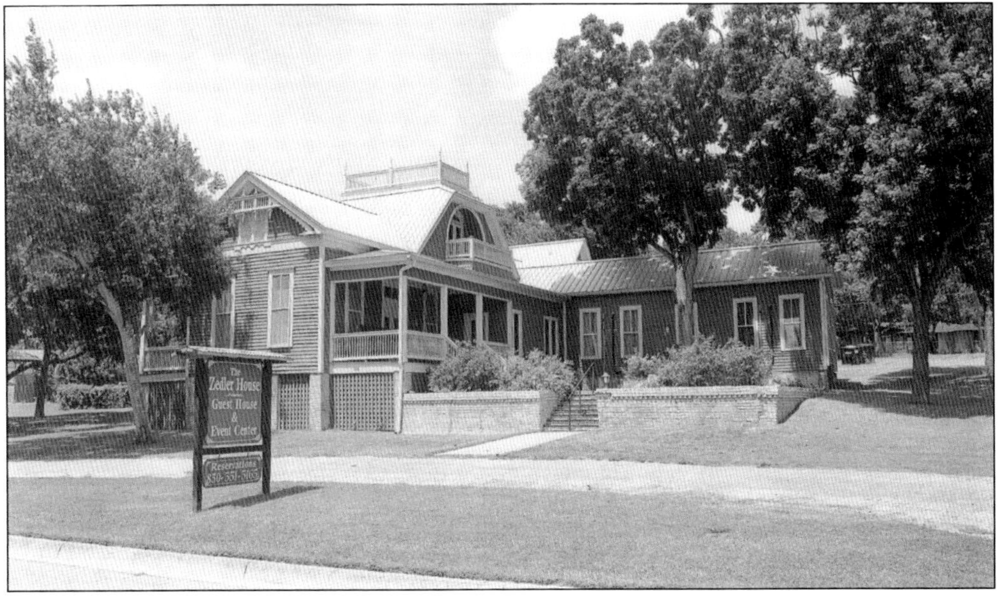

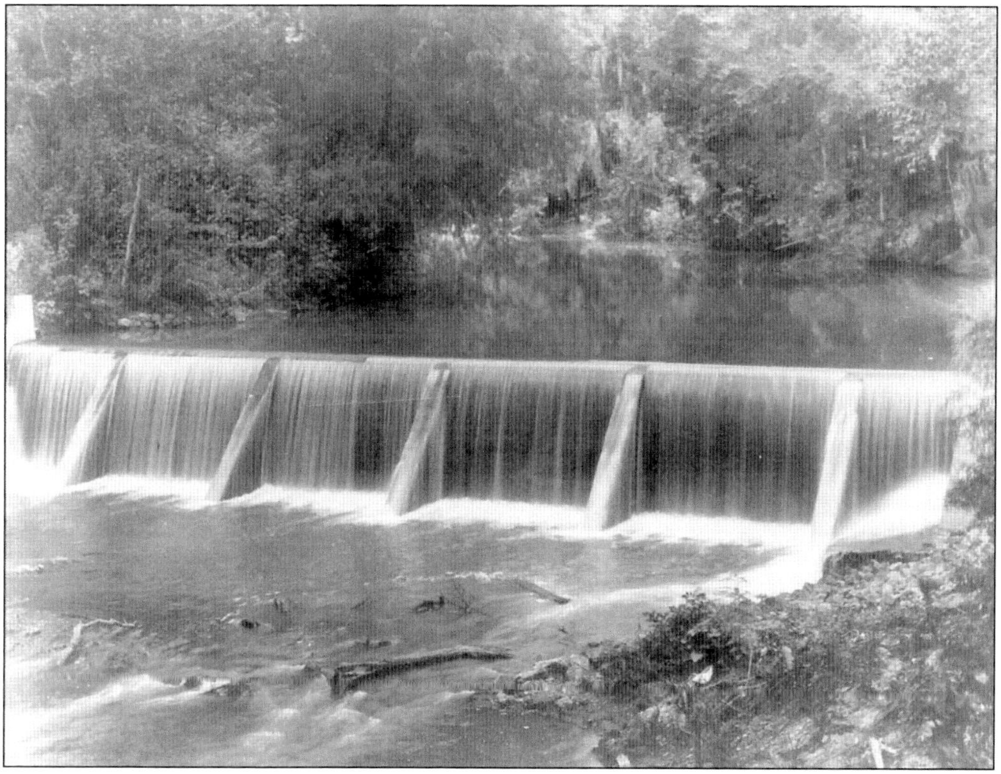

In December 1894, Berthold Zedler, son of Fritz Zedler, purchased land on the San Marcos River at Ottine (ah-TEEN), Gonzales County. Here he would realize his dream of building his own milling industry. Zedler was impressed with the beauty of the area and the waterpower possibilities. After completing his first dam in 1894, Zedler developed a gristmill, sawmill, and cotton gin. In 1910, to increase production at the mills, Zedler built the 12-foot concrete dam above. The photograph below shows bales of cotton produced at Zedler's gin. Prior to the construction of the 1910 dam, cotton production averaged 2,000 bales a year. After completion of the new dam, cotton production grew to a record of 3,500 bales per year. (Both, Zedler Mill Foundation, Luling.)

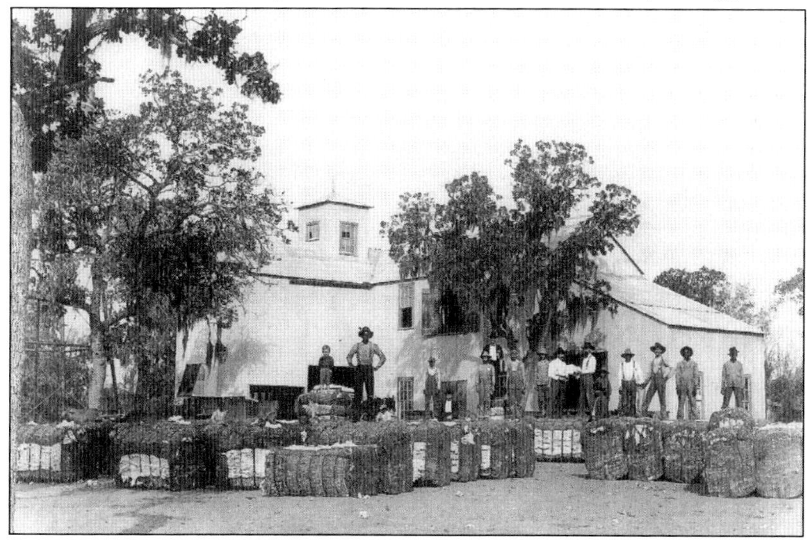

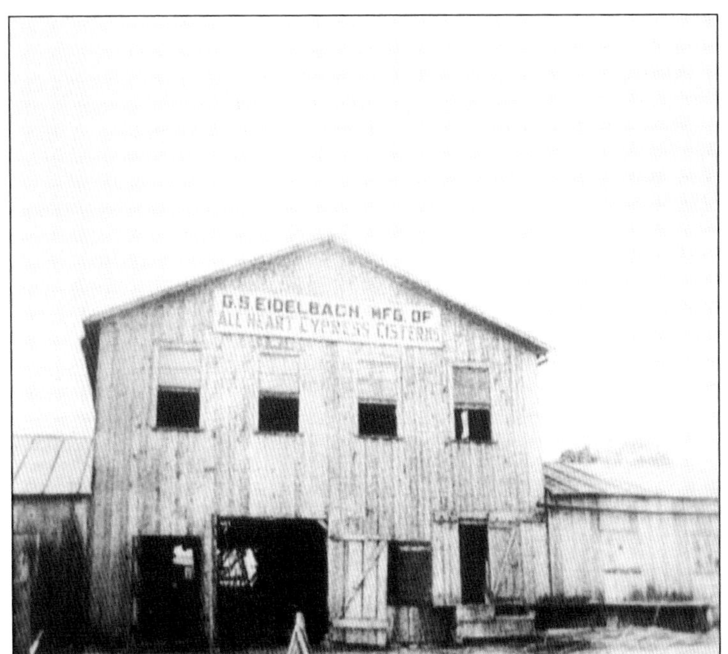

The Eidelbach factory and mill, in Flatonia, Fayette County, manufactured all-heart cypress cisterns and operated as a custom gristmill. When A. Eidelbach started the factory, he made cypress cisterns and oil tanks. Oil tanks were used for the storage of cottonseed oil. When son G.S. took over the business in 1906, he enlarged it by making brewery tanks and kegs along with the cisterns and oil tanks. (Dona Stalmach Carrier.)

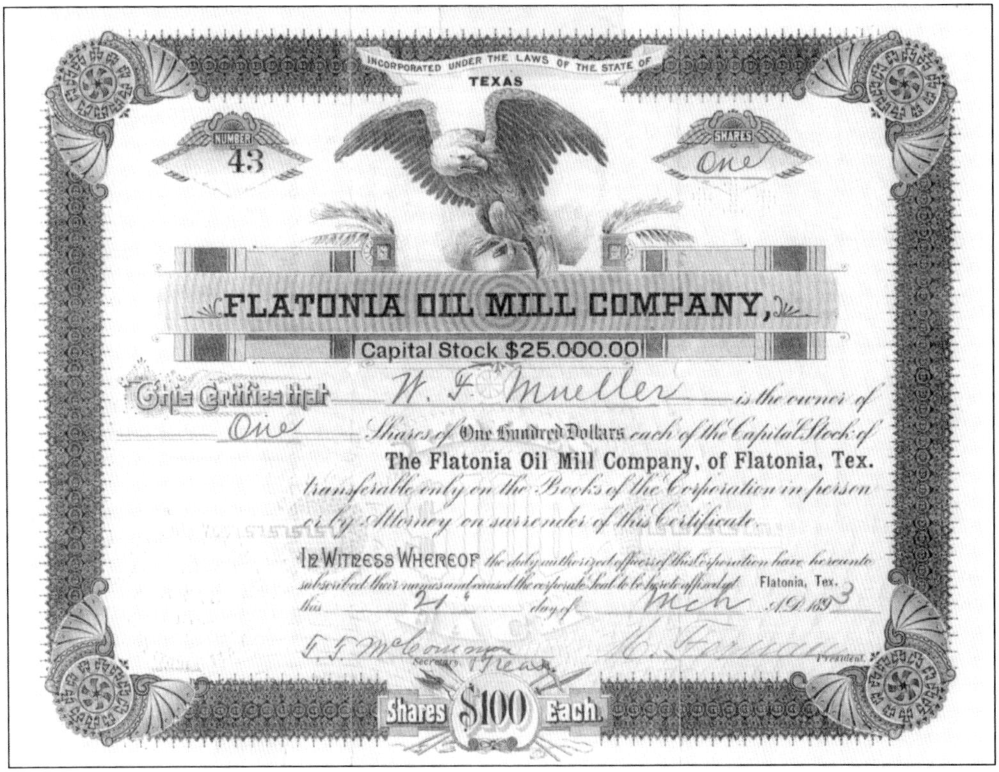

INCORPORATED UNDER THE LAWS OF THE STATE OF TEXAS

NUMBER 43

SHARES One

FLATONIA OIL MILL COMPANY,

Capital Stock $25.000.00

This Certifies that W. F. Mueller is the owner of One Shares of One Hundred Dollars each of the Capital Stock of The Flatonia Oil Mill Company, of Flatonia, Tex. transferable only on the Books of the Corporation in person or by Attorney on surrender of this Certificate.

In Witness Whereof the duly authorized officers of this Corporation have hereunto subscribed their names and caused the corporate Seal to be hereto affixed at Flatonia, Tex. this 21st day of Mch A.D. 1893

Secretary. Treas. President.

Shares $100 Each.

This stock certificate was issued by the Flatonia Oil Mill Company to W.F. Mueller on March 21, 1893. Fayette County and adjoining counties were devoted to the production of cotton. With its easy access and ample railroad facilities, Flatonia Oil Mill Company soon became a leading mill in the production of cottonseed oil. (E.A. Arnim Archives & Museum of Flatonia.)

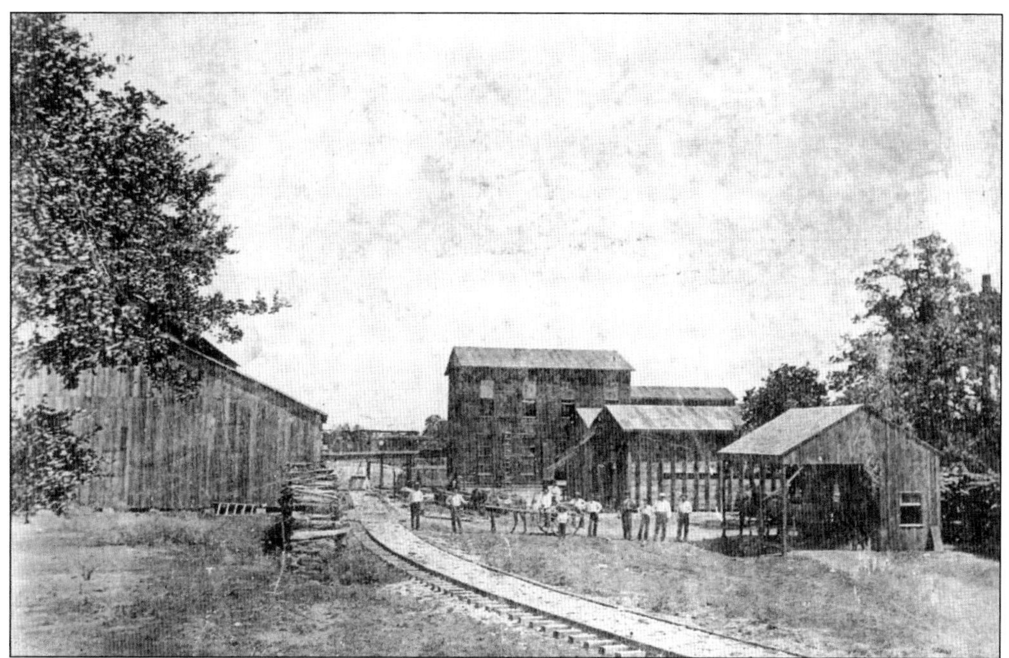

The Flatonia Oil Mill Company, shown in both photographs, was built in 1892 to convert cottonseed into a marketable commodity. Cottonseed was considered as waste until it was discovered to have value for both animal and human consumption. The Flatonia Oil Mill Company enjoyed a reputation for using first-class, up-to-date equipment to convert the raw seed into useable products. Refined oil from the cottonseed was used to make salad oil, cooking oil, soap, and other household products. The mill operated into the 1960s. (Both, E.A. Arnim Archives & Museum of Flatonia.)

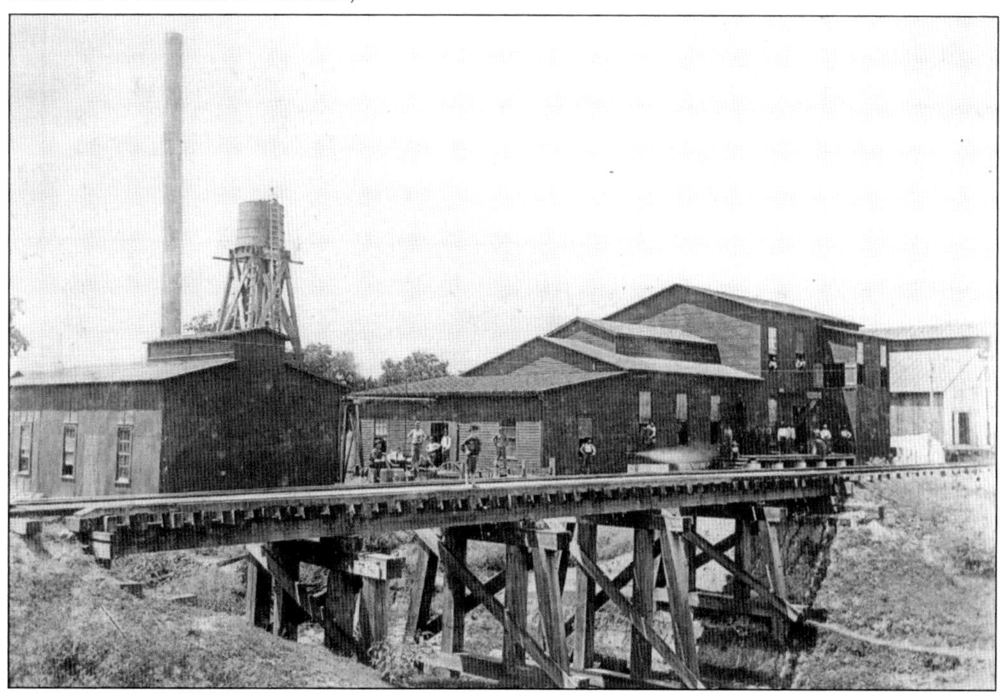

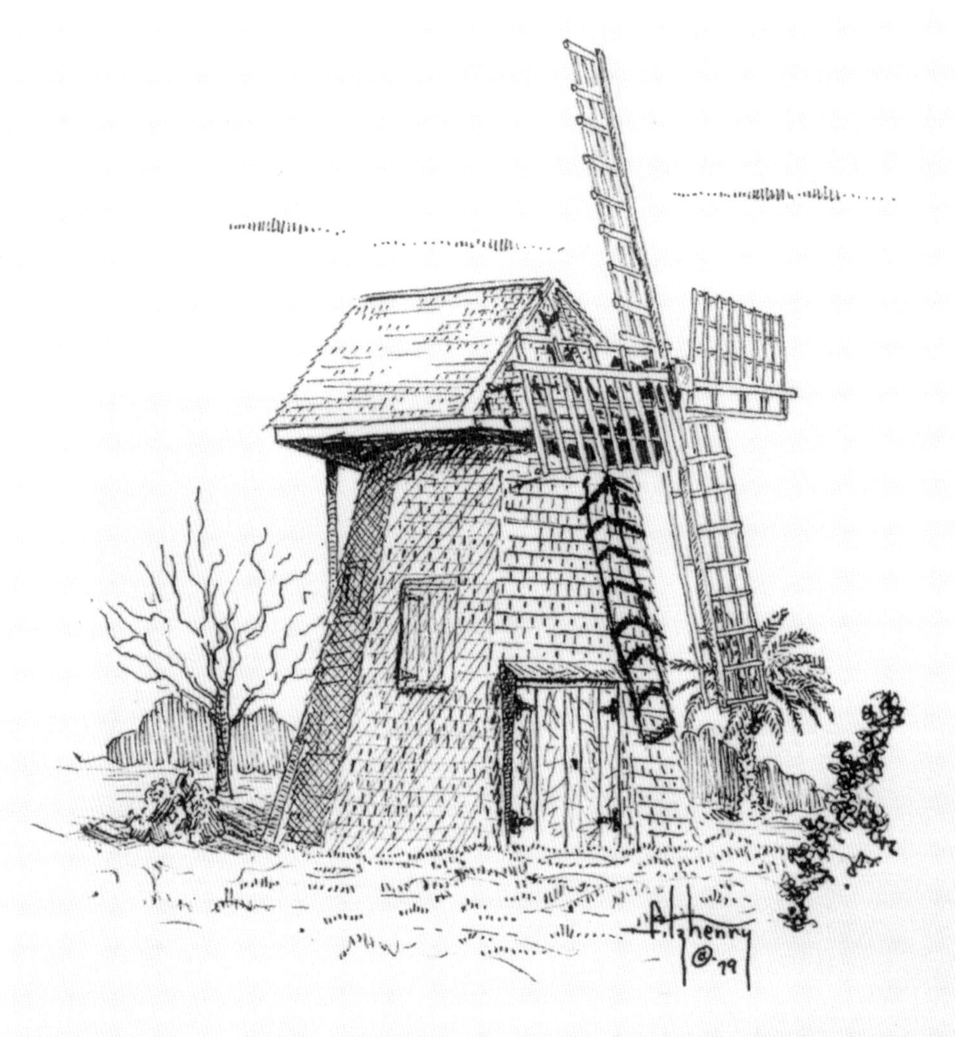

The Old Dutch Windmill at Victoria, Victoria County, was an example of a gristmill powered by wind rather than water. A mill was built before the Civil War east of Goliad, Texas, by E.G. Witte, who had imported the millstones from Europe. These millstones are believed to be one of the earliest sets in the United States to survive. The mill was later sold to Louis Albrecht, who moved it to Coleto Creek, west of Victoria. In 1870, Fred Meiss Jr. purchased the mill and, with the help of Otto Fiek, rebuilt it on Spring Creek, north of Victoria, where it could catch the coastal winds. The windmill stood 35 feet tall and supported four blades approximately 15 feet long. Huge gears, 20 feet in diameter, drove gears attached to a wooden shaft, which in turn moved the grinding stone. In 1935, the Meiss family donated the mill to the Morning Study Club of Victoria, and for 80 years it stood in Victoria Memorial Park. The mill is undergoing restoration and will be relocated to a more natural setting on land south of Inez, Texas. (Jim Fitzhenry.)

Three

The Colorado River

The Route to Texas

The Colorado River rises in northeastern Dawson County, south of Lubbock, and flows southeast to Matagorda Bay. The Colorado River of Texas is not to be confused with the river of the same name that carved out the Grand Canyon. The Colorado River of Texas is 862 miles long, making it the longest river with both its source and mouth in Texas.

Colorado is a Spanish word meaning "reddish," which more aptly describes the waters of the nearby Brazos River. Since the waters of the Colorado have always been a bluish-green color, historians suggest that the names of the two rivers may have been accidently switched by early mapmakers.

During the years of European settlement, the Colorado River had a significant impact on the development of Texas. It was used as an inland water route by early Anglo American colonists and was one of the sites for the settlement of the Old Three Hundred—those who received land grants in Stephen F. Austin's first colony. A decade later, those settlers fought to liberate Texas from Mexico and become an independent country. During the war with Mexico, the river was used as a transportation route for those escaping the advancing Mexican army.

In 1839, a few years after Texas became a sovereign nation, the Capital Commission of the Republic of Texas chose the spot where the river flows from the Balcones Escarpment as the site of a new capital of the republic. They named it Waterloo; it was eventually renamed Austin.

Many of the early millwrights coming into Texas recognized the power of the waters of the Colorado River and chose spots along the river or one of its tributaries as an ideal location for their gristmills. Mills in this chapter that were built along the Colorado or one its streams include Lange's Mill at Doss, Reliance Mill in Fredericksburg, C.H. Guenther Live Oak Mill near Fredericksburg, San Saba Mill in San Saba County, the Roller Mill and Mormon Mill near Burnet, Chadwick Mill in Lampasas County, and Anderson Mill at Volente near Lake Travis.

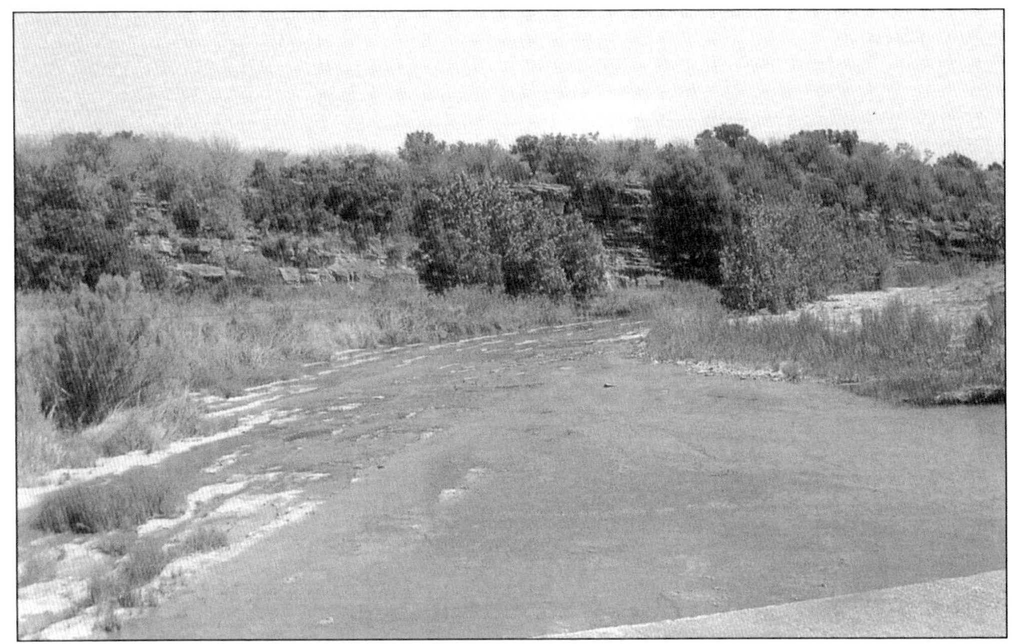

Threadgill Creek, in northwest Gillespie County, is home to Lange's Mill, one of the last and best preserved burrstone mills in Texas. In 1849, brothers John E. and Thomas Doss built a gristmill and distillery on Threadgill Creek, and they later added a dam and sawmill. The distillery and sawmill were washed away in 1854, but the gristmill survived. In 1859, William F. Lange acquired the property and expanded the gristmill's capabilities. (Author's collection.)

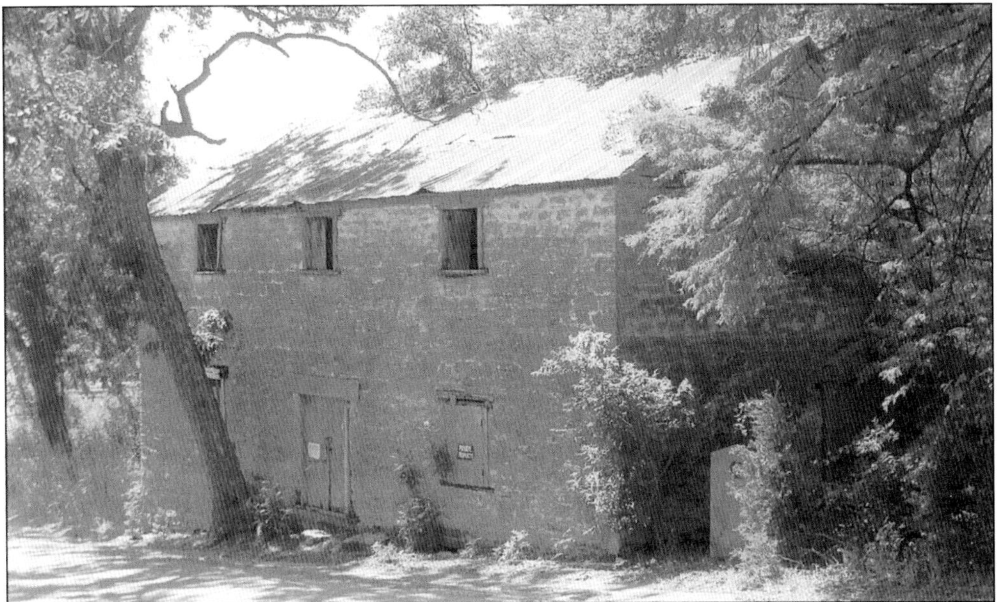

Lange's Mill has been damaged by floods and rebuilt several times. The present three-story stone structure dates from 1875. A visitor to the mill several years ago wrote that stored inside was a horse-drawn buggy probably in use when the mill was at its peak of production. The upper floor was scattered with odds and ends of tools and equipment once used in the mill. All items were mementos of another way of life. (Author's collection.)

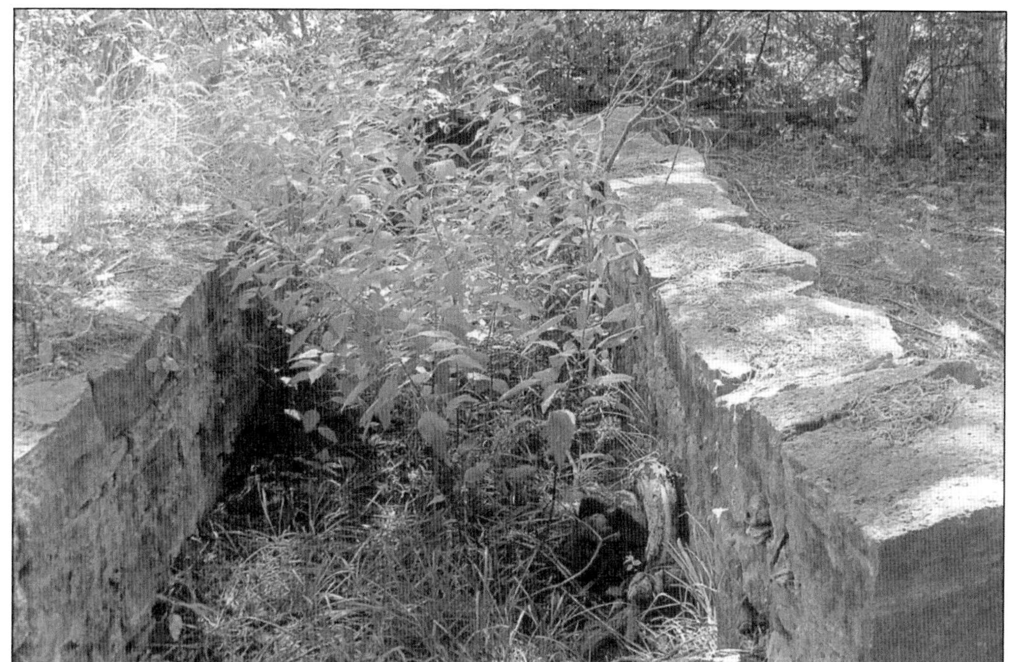

This photograph shows the abandoned millrace at Lange's Mill. This is a typical sight at the old mills. The mill buildings are now in ruins, their walls lie in piles of rubble, their dams are broken by floods, and their millraces are choked with vines and debris. The service of some has been acknowledged by historical markers, while others have been long forgotten, but all are silent observers of the changes that have come over the country. (Author's collection.)

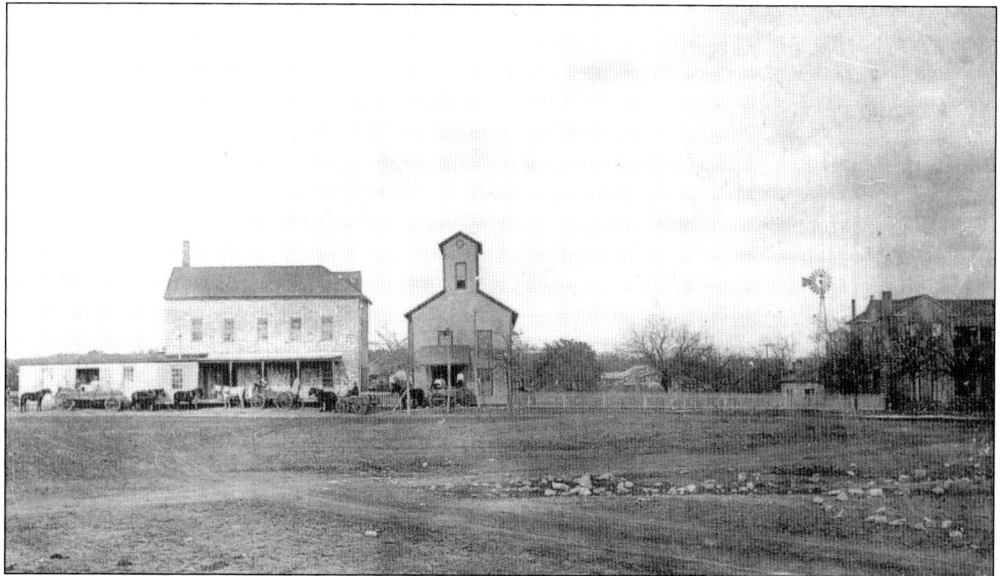

In 1862, brothers Felix and Frank VanDerStucken opened the Reliance Flour Mill in Fredericksburg. In 1889, they renovated the entire complex into a roller mill that produced flour of the highest grade. The photograph gives a view of the open square with the Reliance Flour Mill on the north side near the VanDerStucken house, at far right. (The University of Texas at San Antonio Libraries, Special Collections.)

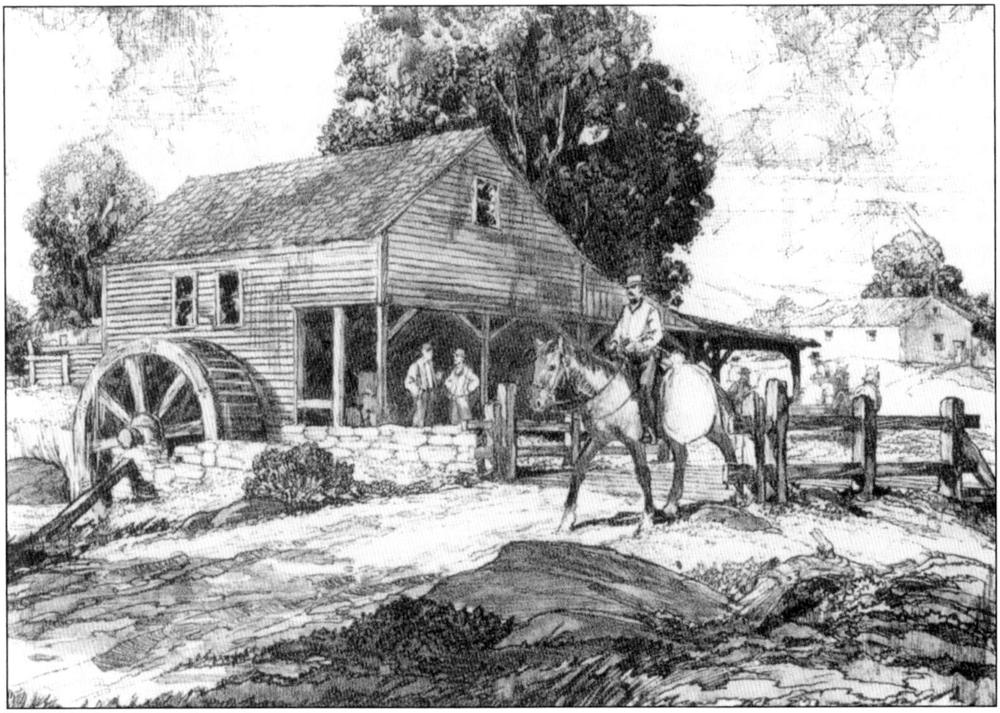

In 1851, German millwright Carl Hilmar Guenther built his first gristmill in Texas about three miles southwest of Fredericksburg, Gillespie County. It was situated at the bend in Live Oak Creek just before it flowed into the Pedernales River. The mill was powered by an undershot wheel, turned by water flowing through a flume leading downward to the wheel's base. This painting was done by Lloyd Harting. (C.H. Guenther & Son, Inc., Archives, all rights reserved.)

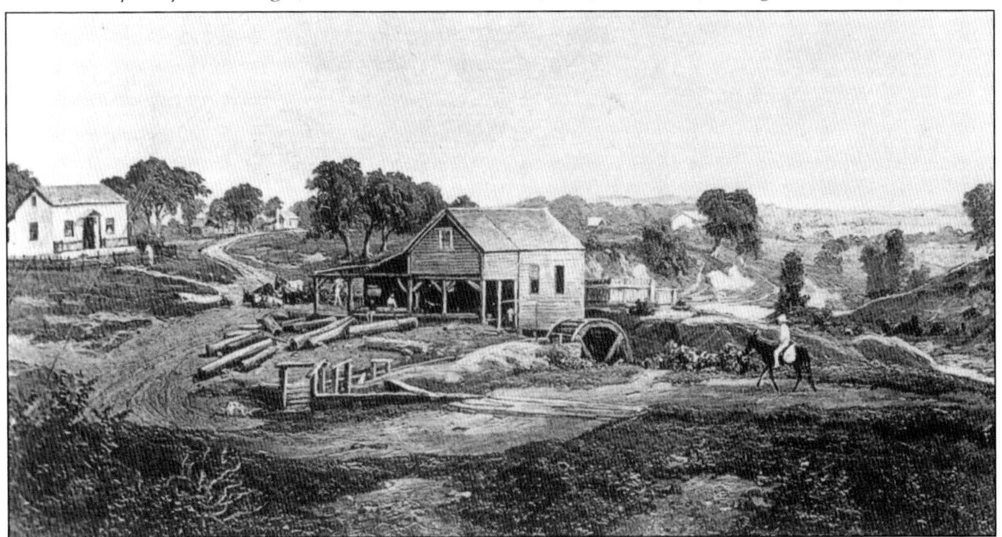

This 1855 painting by Guenther's neighbor Hermann Lungkwitz, a noted Texas landscape artist, shows Guenther's mill with a sawmill added and, on the left, the stone house which Guenther built for his new bride. Guenther operated his mill on Live Oak Creek until the late 1850s before selling the mill operation to his father-in-law and moving to the bustling city of San Antonio. (C.H. Guenther & Son, Inc., Archives, all rights reserved.)

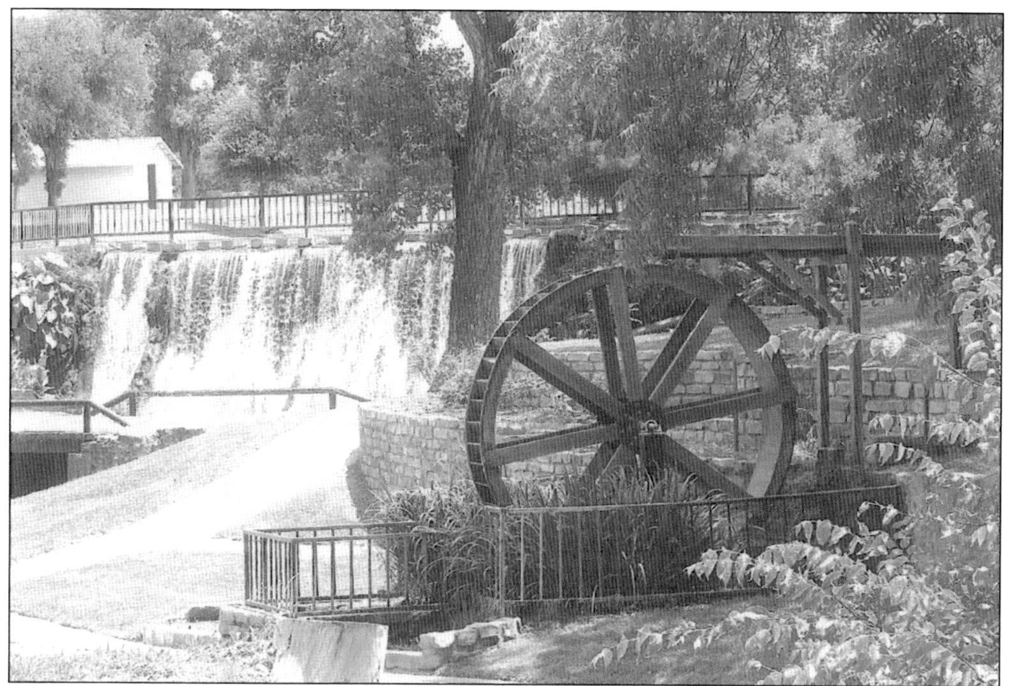

The Colorado and San Saba Rivers have been a source of water for the citizens of San Saba since its earliest days. In 1867, John "Shorty" Humphrey Brown, a founder of San Saba, constructed a gristmill on Mill Creek in an area now known as Mill Pond Park. The mill was run by water from the San Saba Springs. Brown also constructed a building that housed the machinery for operating the gristmill and a cotton gin. This building, shown below, became known as Mill Pond House. After the mill ceased operating, the city's water system machinery was housed in the building. When the area was established as a park, the San Saba Garden Club restored the building in 1956, and it now serves as their meeting place. (Above, Lynn Blankenship; below, author's collection.)

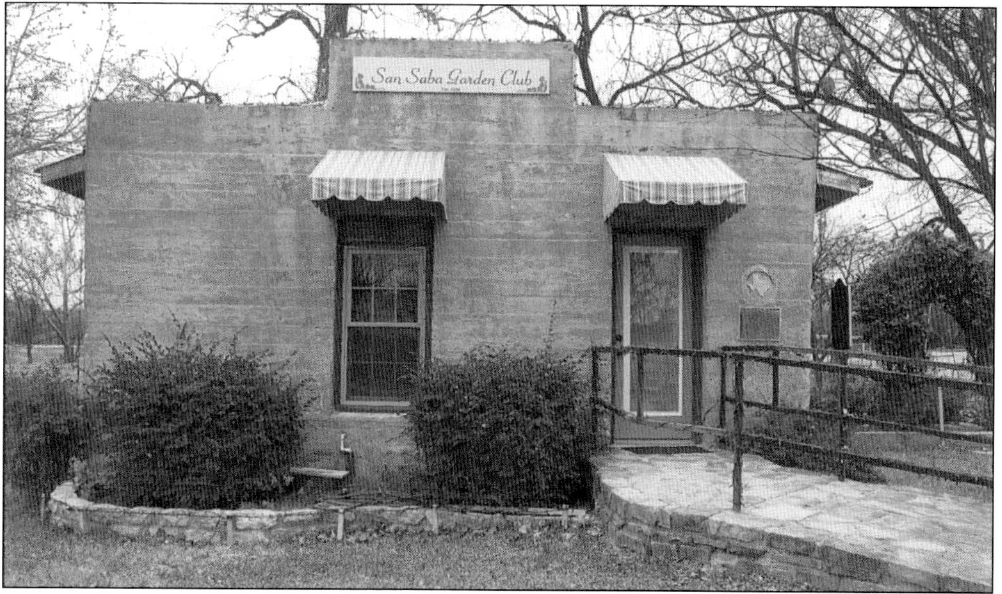

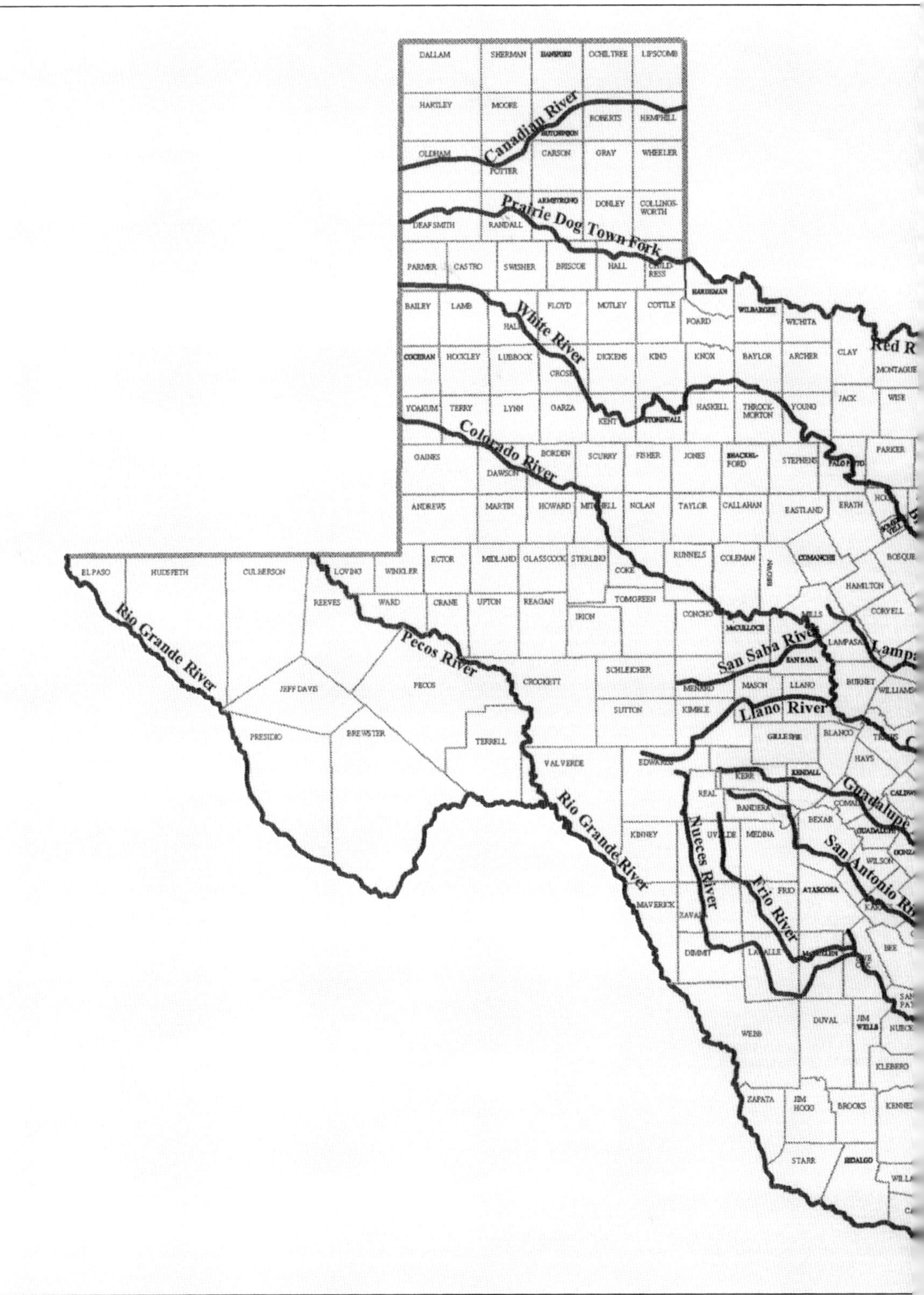

DALLAM
SHERMAN
HANSFORD
OCHILTREE
LIPSCOMB
HARTLEY
MOORE
Canadian River
ROBERTS
HEMPHILL
HUTCHINSON
OLDHAM
CARSON
GRAY
WHEELER
POTTER
ARMSTRONG
DONLEY
COLLINGS-
WORTH
Prairie Dog Town Fork
DEAF SMITH
RANDALL
PARMER
CASTRO
SWISHER
BRISCOE
HALL
CHILD-
RESS
HARDEMAN
BAILEY
LAMB
FLOYD
MOTLEY
COTTLE
FOARD
WILBARGER
WICHITA
White River
HALE
CLAY
COCHRAN
HOCKLEY
LUBBOCK
DICKENS
KING
KNOX
BAYLOR
ARCHER
MONTAGUE
CROSBY
YOAKUM
TERRY
LYNN
GARZA
KENT
STONEWALL
HASKELL
THROCK-
MORTON
YOUNG
JACK
WISE
Colorado River
GAINES
DAWSON
BORDEN
SCURRY
FISHER
JONES
SHACKEL-
FORD
STEPHENS
PALO PINTO
PARKER
ANDREWS
MARTIN
HOWARD
MITCHELL
NOLAN
TAYLOR
CALLAHAN
EASTLAND
ERATH
EL PASO
HUDSPETH
CULBERSON
LOVING
WINKLER
ECTOR
MIDLAND
GLASSCOCK
STERLING
COKE
RUNNELS
COLEMAN
BROWN
COMANCHE
BOSQUE
HAMILTON
REEVES
WARD
CRANE
UPTON
REAGAN
TOM GREEN
IRION
CONCHO
McCULLOCH
MILLS
CORYELL
Rio Grande River
Pecos River
San Saba River
SAN SABA
LAMPASAS
JEFF DAVIS
PECOS
CROCKETT
SCHLEICHER
MENARD
MASON
LLANO
BURNET
SUTTON
KIMBLE
Llano River
PRESIDIO
BREWSTER
TERRELL
GILLESPIE
BLANCO
HAYS
VAL VERDE
EDWARDS
KERR
KENDALL
REAL
BANDERA
BEXAR
Guadalupe
Rio Grande River
KINNEY
UVALDE
MEDINA
Nueces River
San Antonio
WILSON
Frio River
FRIO
ATASCOSA
MAVERICK
ZAVALA
DIMMIT
LA SALLE
McMULLEN
WEBB
DUVAL
JIM WELLS
KLEBERG
ZAPATA
JIM HOGG
BROOKS
STARR
HIDALGO

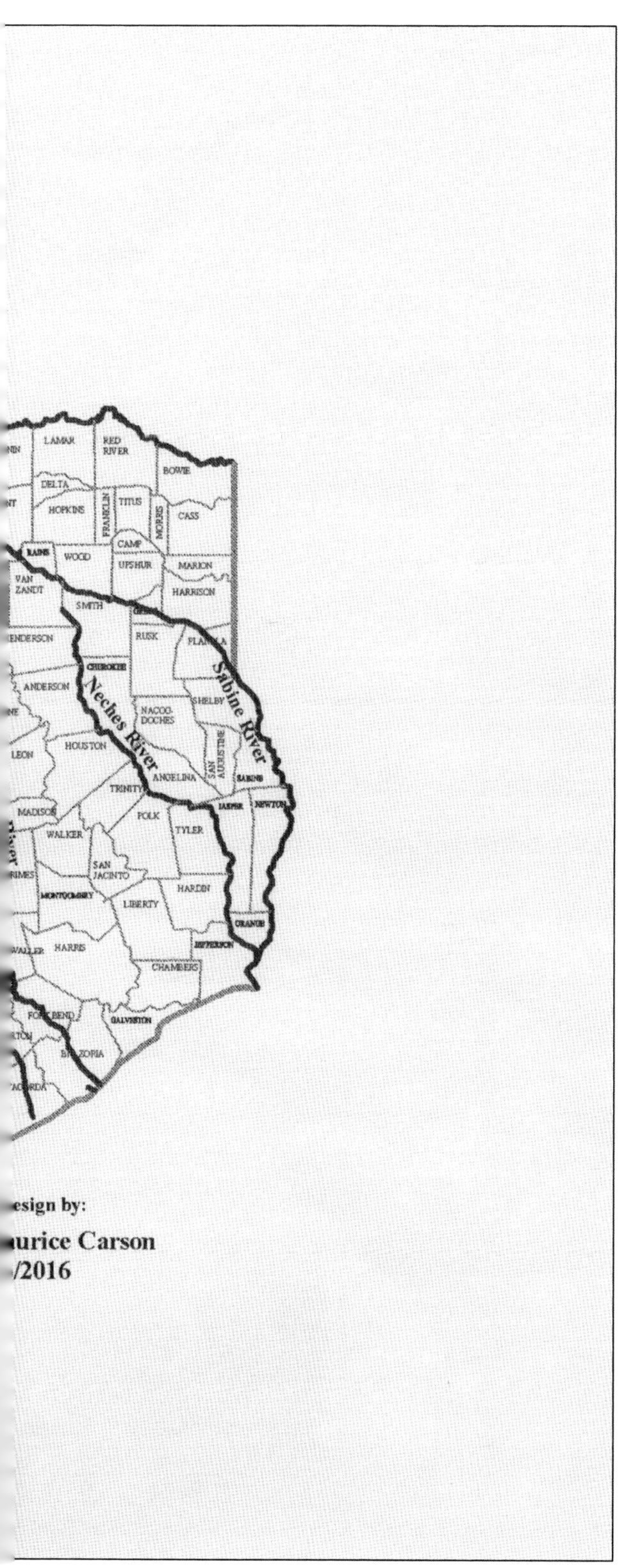

The rivers of Texas were hosts to hundreds of gristmills. When settlers were looking for a mill site, they often looked at the springs and the streams that branched from the rivers. Springs usually provided a more reliable source of water even in the dry times of summer. Streams were less likely to flood than rivers, thus preserving the life of the mill. Most of the millwrights who immigrated to Texas came with milling experience, and as soon as they found a suitable site, they established a mill. Some of those mill settlements grew into towns that eventually became some of the most populous and productive cities in the state. (Maurice Carson.)

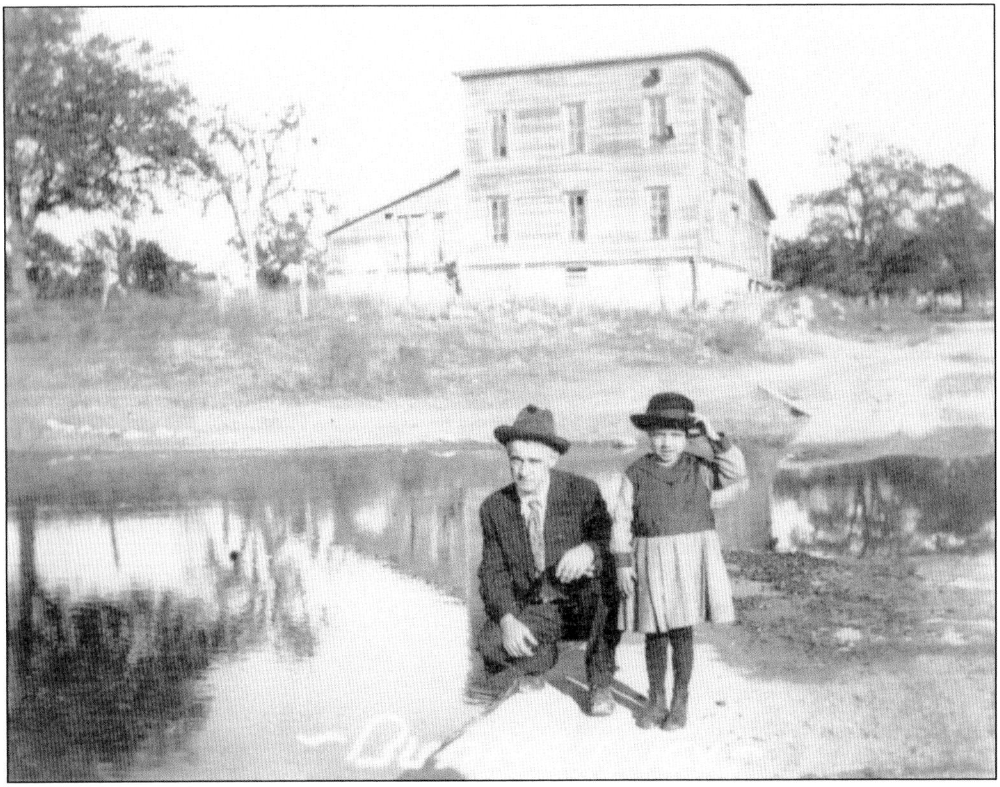

The Roller Flour Mill on Hamilton Creek, Burnet County, was powered by steam generated by a large boiler and was considered a modern, efficient operation. The boiler exploded in 1880, killing the owner's father, Bryce Smart; a worker, H.H. Hall; and Johnny Blackburn, a 13-year-old-boy who had stopped at the mill to watch the machinery. John Allen (Al) Kinkead and his daughter Grace Elizabeth are pictured with the mill. (Herman Brown Free Library, Burnet.)

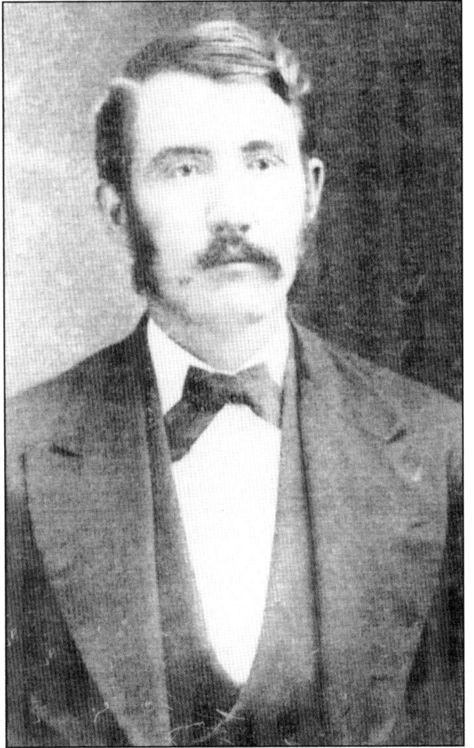

John White Smart, owner of the Roller Flour Mill, rebuilt and continued to operate the mill until he sold it a few years later. It was razed in 1920. Smart came to Texas from Missouri around 1851. Prior to coming to Burnet County, Smart operated a gristmill in Williamson County and a tanning yard that made shoes for Confederate troops. Smart enlisted as a Confederate private and served until the war was over. (Herman Brown Free Library, Burnet.)

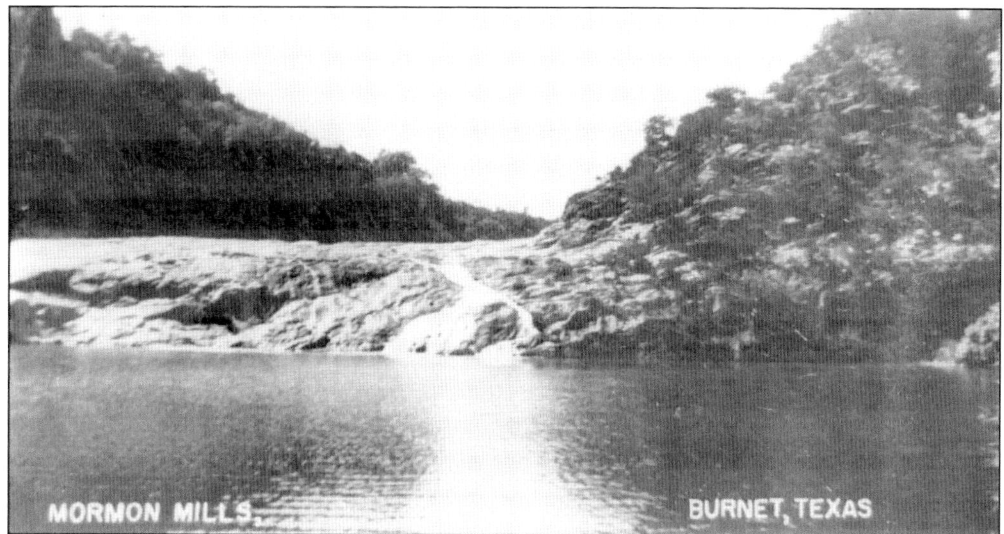

In early 1851, twenty Mormon families settled near Hamilton Creek, Burnet County, where the creek cascaded into a 60-foot-deep pool. Smart's Roller Mill was located on the northern end of the creek near Burnet. The Mormon Mill was located on the southern end of the creek near Marble Falls. Hamilton Creek flowed year-round, providing power for mills that served the Burnet, Marble Falls, and Austin areas. (Herman Brown Free Library, Burnet.)

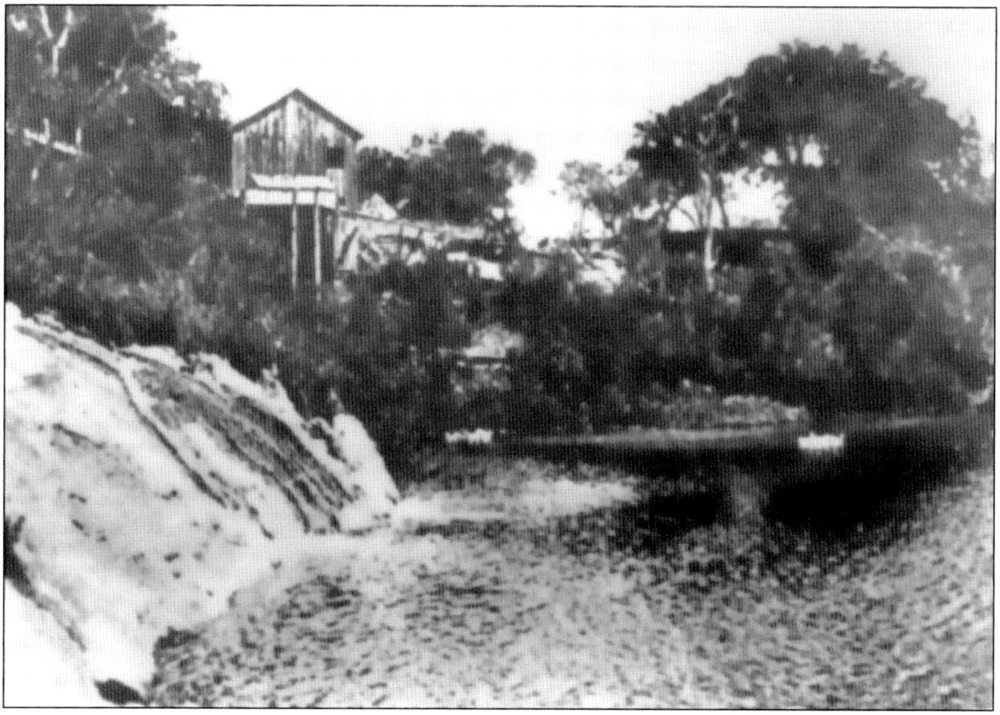

The Mormon Mill was a three-story frame structure with a 26-foot overshot wheel. Next to the main building was a smaller building, with the two buildings connected by a gangway. Under the leadership of Lyman Wight, a colonist of 250 people set up the mill, which was used as a flour and gristmill, a lumber mill, a cotton gin, and a furniture factory. In 1853, Wight sold the property to Noah Smithwick. (Herman Brown Free Library, Burnet.)

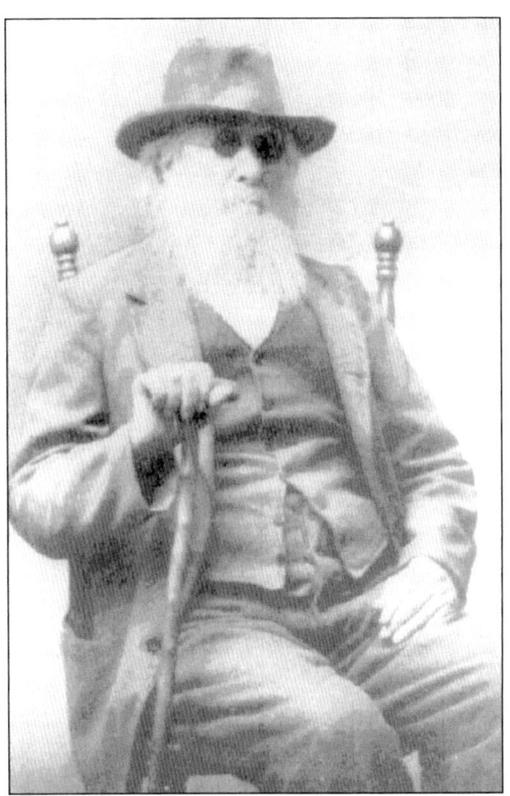

When Noah Smithwick took ownership of the Mormon Mill, he converted it to a flour mill only, thus prompting local farmers to raise more wheat. The mill operated until 1901, when the owner, Price Kinser, had it demolished and used the lumber to build a barn. Only the Mormon cemetery remains. The state erected a historical marker at the mill site in 1936. (Herman Brown Free Library, Burnet.)

This marker, with a millstone background, commemorates Chadwick Mill, which was built on the Colorado River in western Lampasas County. The mill was built in 1874–1875 by Henry A. Chadwick and son James Milam. By 1900, the settlement that grew up around the mill took the name Chadwick. The mill and townsite were located three miles north of the marker, which sits at Highway 190 at the Colorado River Bridge near San Saba. (Author's collection.)

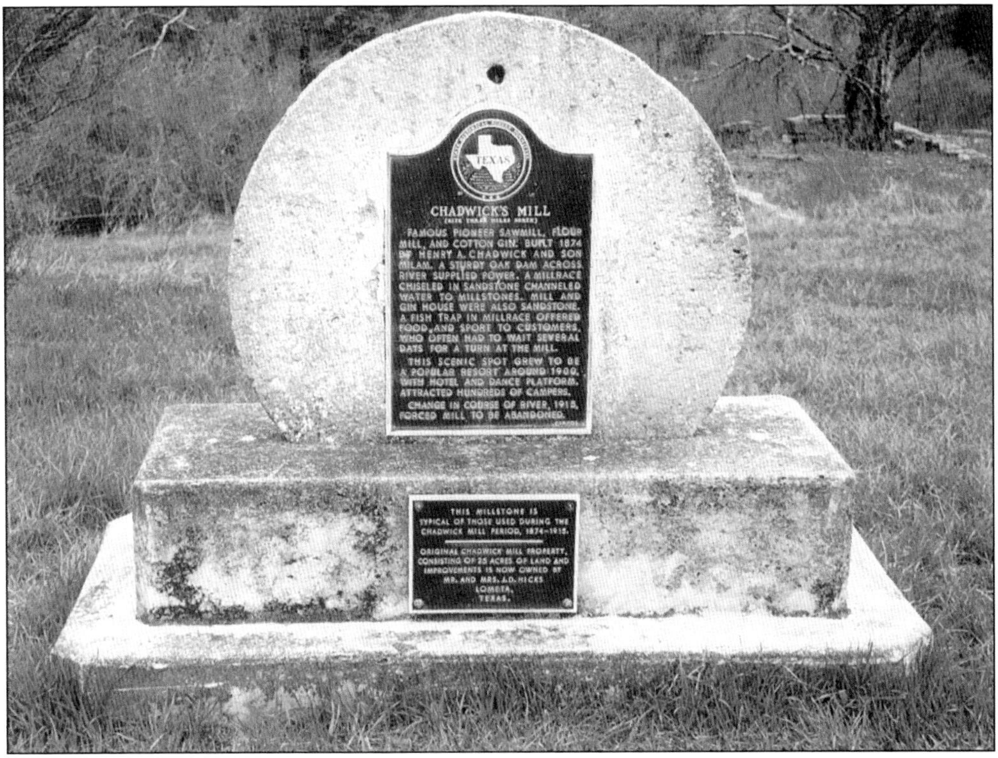

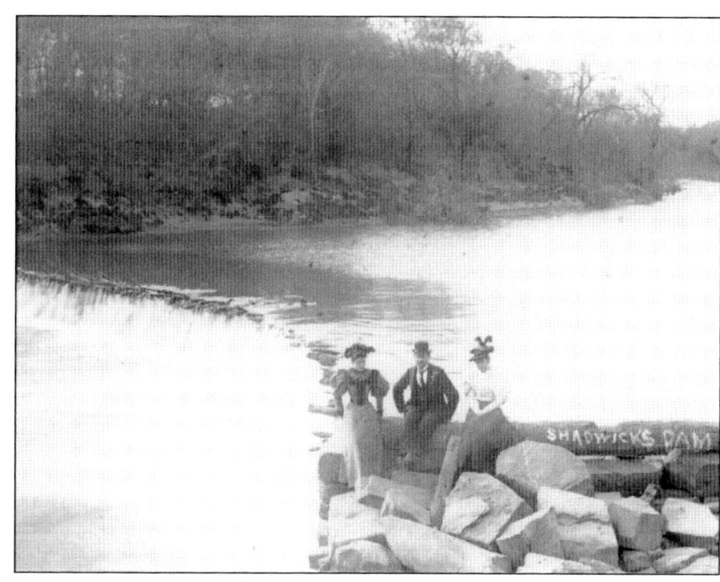

Chadwick Mill's dam was constructed out of stone and burr oak logs floated to the site from upstream. Burr oak was selected because it would never rot or deteriorate as long as it was kept under water. The dam was made leak-proof by sediment building up after rocks and cedar were piled on the upstream side. The Sanderson family pose at the dam for this early-1900s photograph. (San Saba County Historical Museum.)

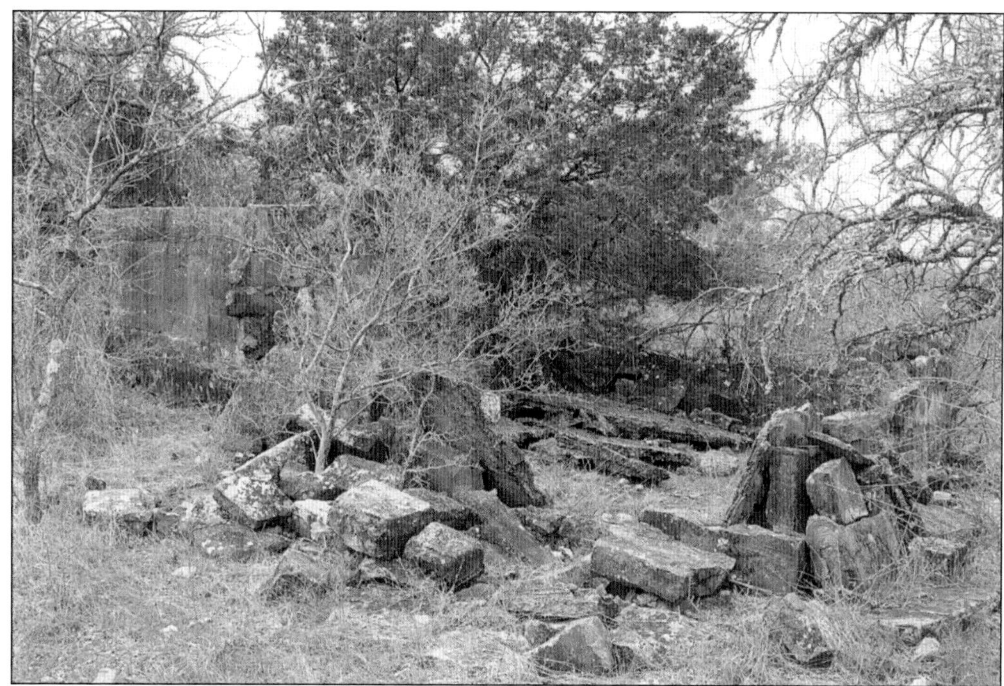

This photograph shows the ruins of one of the businesses that were once a part of the Chadwick community. After Chadwick sold the mill in 1904, several different owners continued its operation. A post office was established in 1910 but was discontinued in 1918. The mill ceased operation in 1915 due to flooding of the Colorado River. All that remains of this once-thriving community is the Chadwick Cemetery, which is still in use. (Author's collection.)

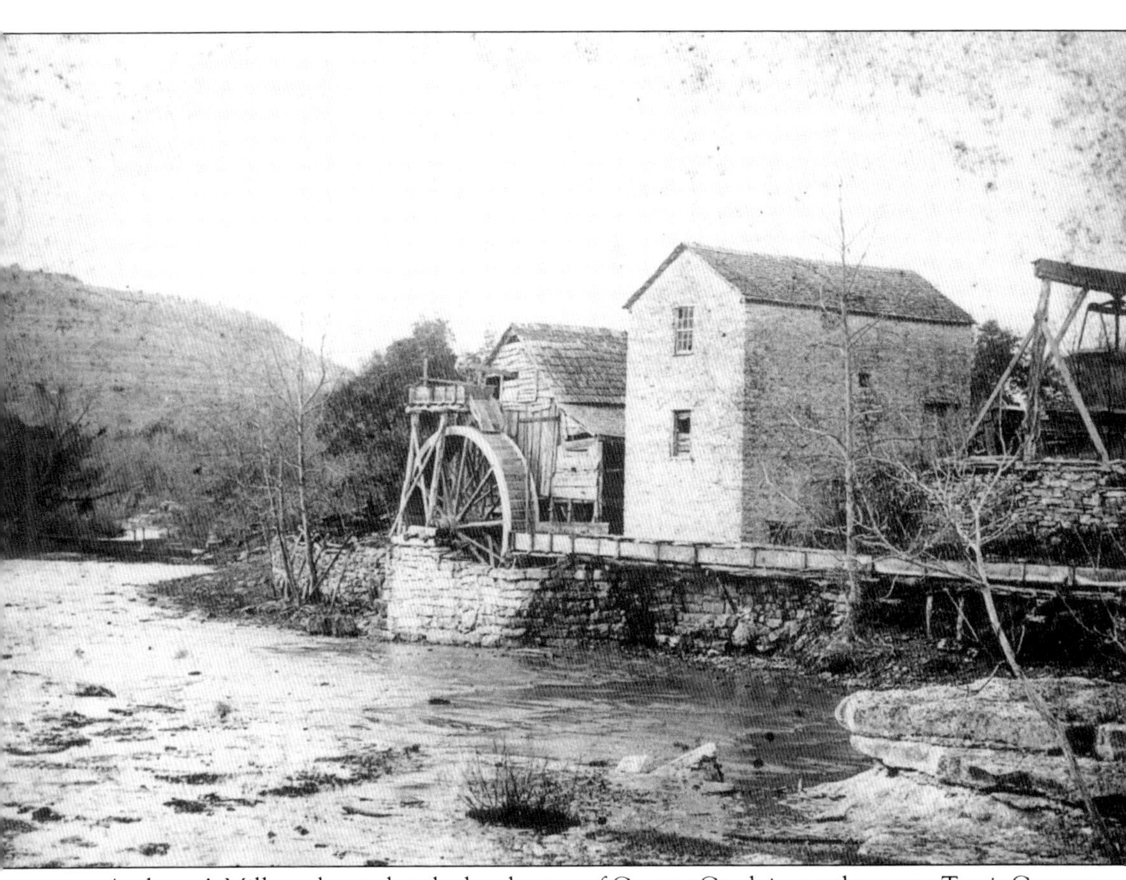

Anderson's Mill was located at the headwaters of Cypress Creek in northwestern Travis County, about 16 miles northwest of Austin. Thomas Anderson and his two sons, Abe and Ed, built the gristmill in the early 1860s. The 18-foot cypress wheel was driven by undershot water power. Early in 1863, under the direction of the Texas State Military Board, the mill was converted to manufacture gunpowder for the Confederacy. When Anderson found that the undershot wheel did not provide sufficient power, an overshot flume was added. The tall flume was built from the mill pond to the top of the wheel so that both the upper and lower millrace were in use. When the war was over, Anderson converted the mill back to a gristmill, and in 1870, he added cotton gin machinery. The apparatus to the right of the mill is a cotton press, which was used to form the cotton into bales. (Alvin Thomas Jackson Papers, di_10680, the Dolph Briscoe Center for American History, the University of Texas at Austin.)

Thomas Anderson, born in Virginia in 1820, came to the Texas frontier well prepared to be a millwright. He had served a 10-year apprenticeship in Virginia learning and developing the skills of a millwright. He was a designer and an engineer with a mastery of both civil and mechanical disciplines. He had experience building dams, flumes, and aqueducts. In addition, he was a well-respected businessman. (Anderson Mill Memoria.)

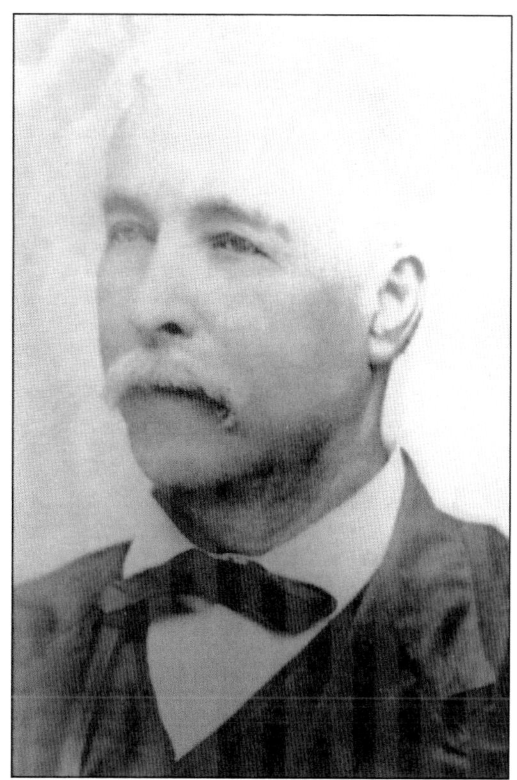

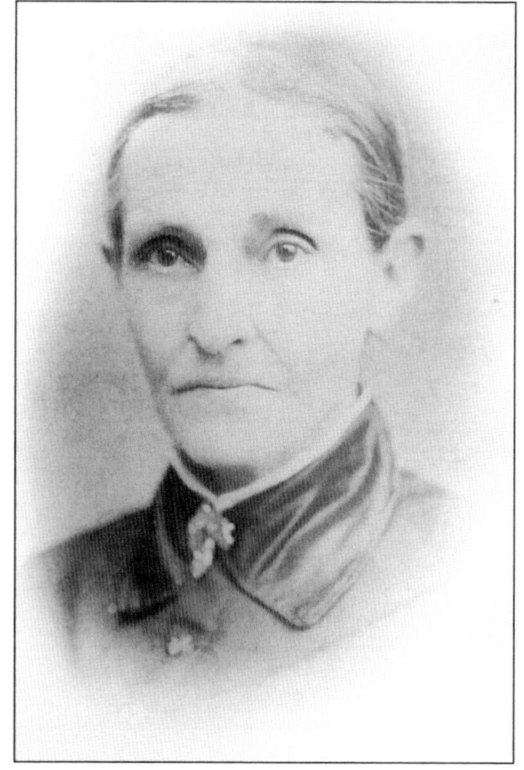

In the early 1850s, Lucy Anderson, wife of Thomas Anderson, left a comfortable plantation life to come by oxen from Pennsylvania, by way of Virginia, to the new state of Texas along with her husband and four children. Lucy was a gracious hostess to the many farmers and their families camped in her front yard waiting their turn to have their corn ground. Some had to wait several days. (Anderson Mill Memoria.)

Locations of Original Anderson Mill and Replica

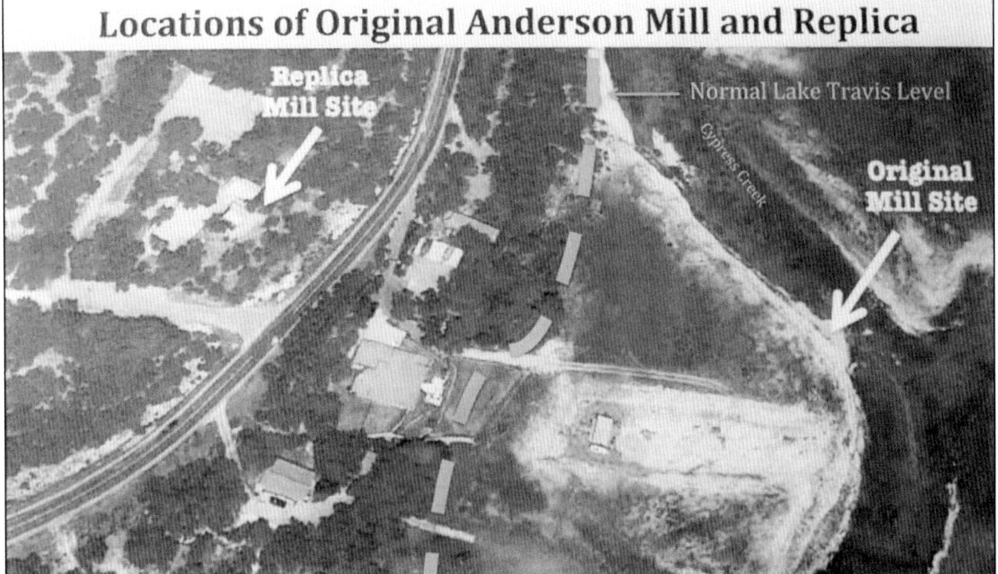

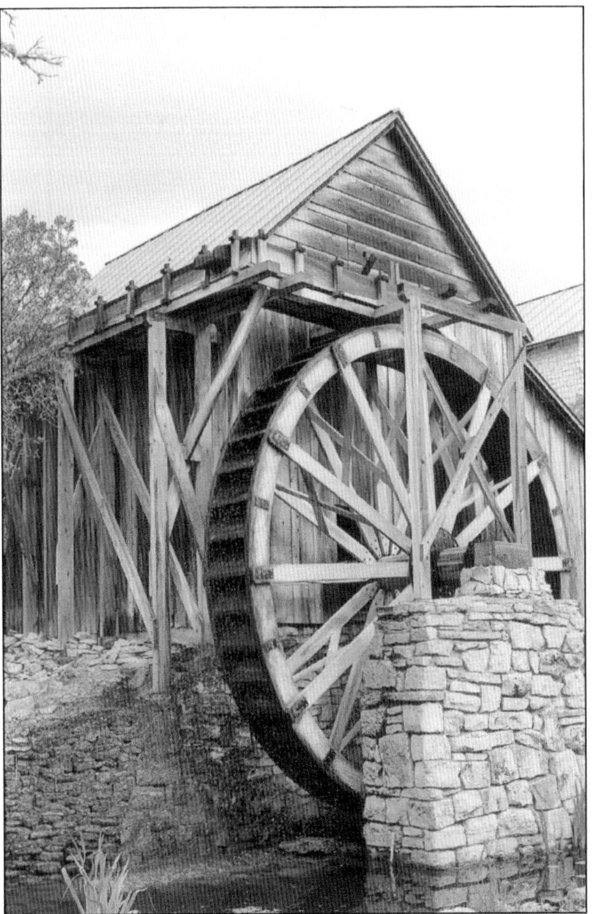

After Thomas Anderson's death on July 23, 1894, the family moved to Austin. The mill equipment was sold and the structure deserted. In 1941, the ruins of the mill were torn down and stones carted away in preparation for the construction of Lake Travis. The site of the old mill now lies beneath the waters of the lake. (Anderson Mill Memoria.)

The memory of the mill will live on due to the efforts of the Anderson Mill Garden Club members who worked for several years to earn enough money to reconstruct the mill. The restoration work started at their clubhouse, Robinson Hall, located on a hill above the former site of the mill. This photograph focuses on the restored waterwheel. (Author's collection.)

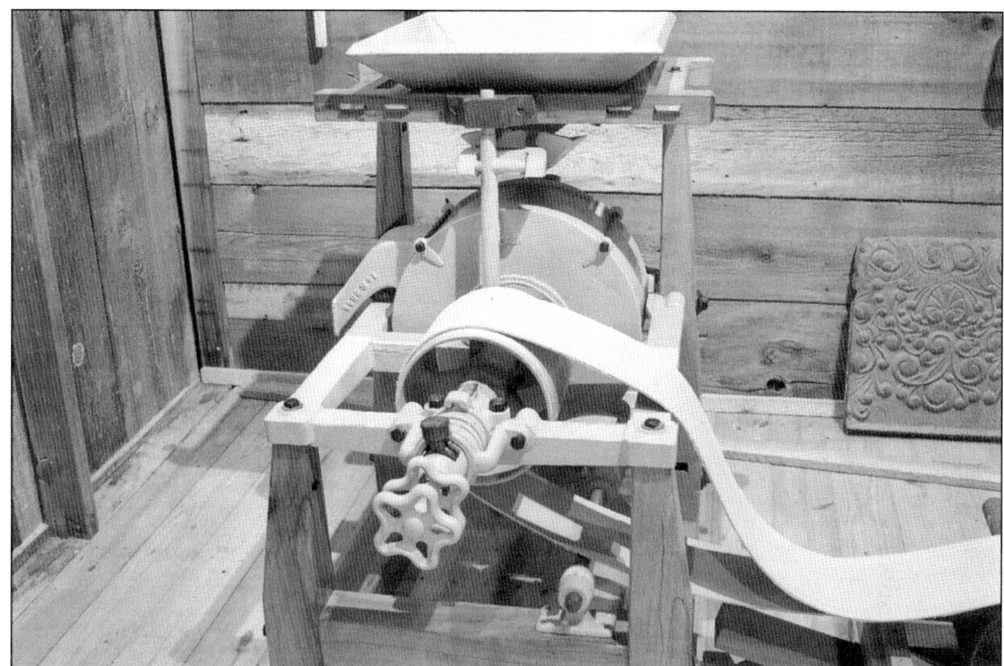

This small, fully restored vertical mill, located inside the millhouse, is thought to be similar to one in the original Anderson's Mill and is dated from the 1920s. The metal parts are original, but all parts made of wood have been rebuilt. The rebuilt flat wooden pulley below is mounted to the outside waterwheel by a shaft. The shaft was often a hewn log. The wooden slats covering the shaft were added to lend a degree of authenticity. When the mill is operating, the flat belt provides the connection between the mill and the waterwheel, and by increasing the flow of water over the waterwheel, sufficient power is generated to operate the mill. (Both, author's collection.)

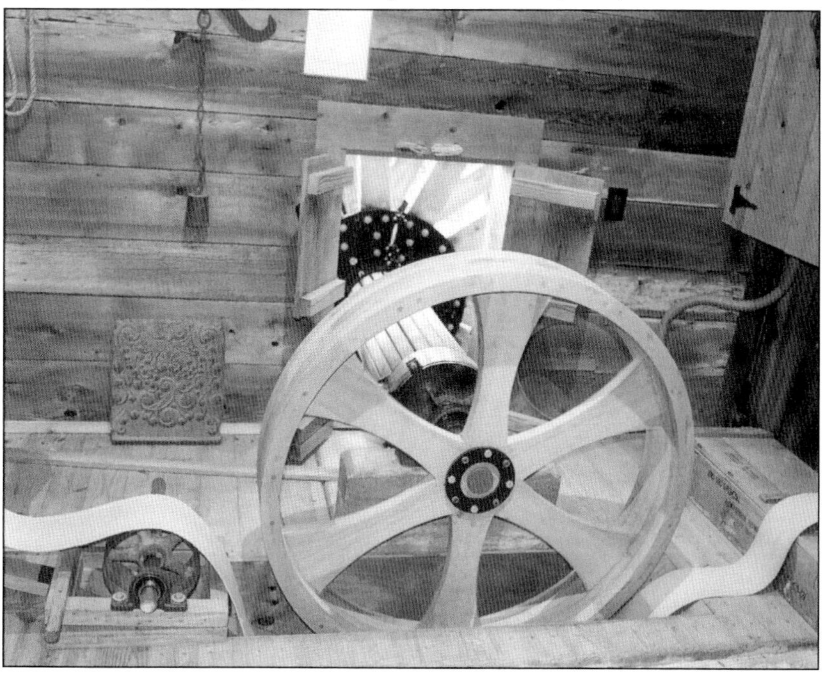

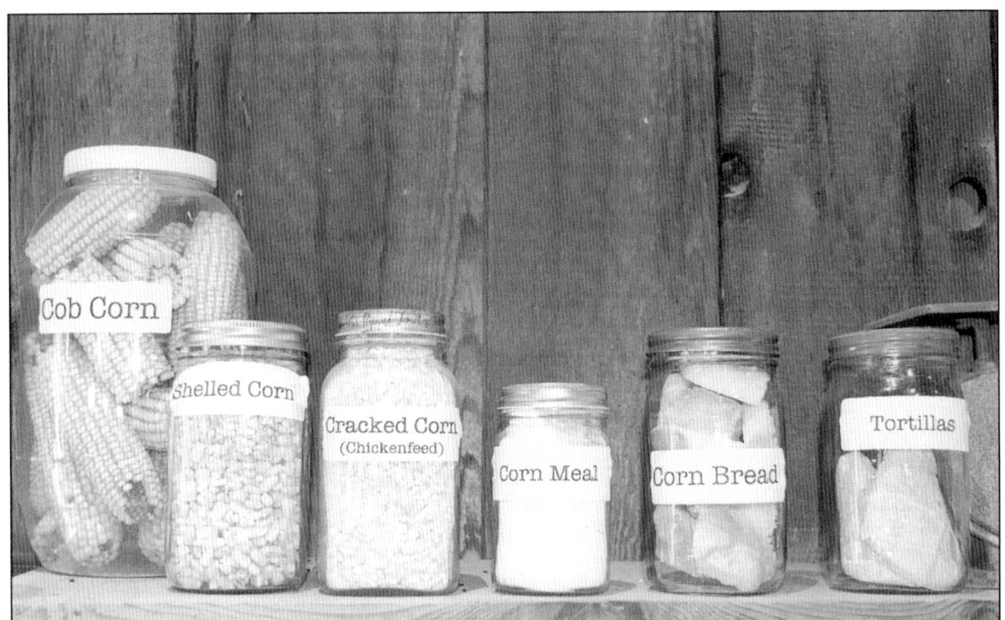

This display illustrates the steps involved in turning harvested corn-on-the-cob into more refined products. Normally, the farmer would bring shelled corn to the mill for grinding. The corn could be ground into cracked corn, which is used for chicken feed, or it could be ground into cornmeal, which is used for making breads and other foods for human consumption. (Author's collection.)

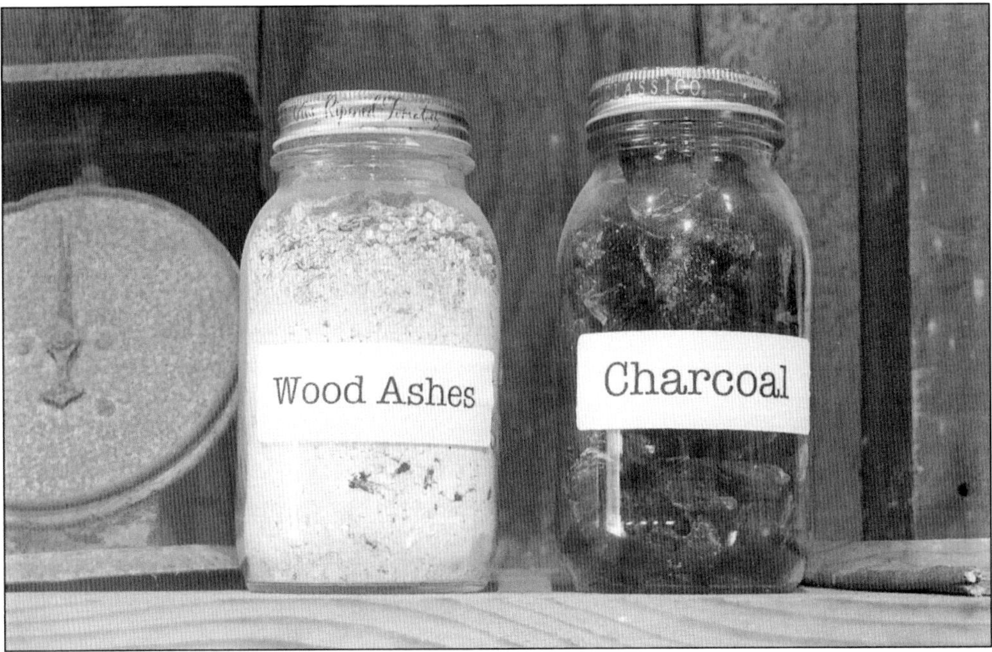

The bat-infested caves in the area made the mill an ideal place for making gunpowder. Bat guano from the caves was placed in vats with layers of ashes and broom weed. Water was added to the top and piped off the bottom after it had filtered down through the vats. The liquid was then heated until saltpeter crystallized. This saltpeter was then ground and mixed with sulfur and charcoal to form gunpowder. (Author's collection.)

This is the third replica waterwheel to be constructed at the mill. The first one was constructed and installed in 1965, the second one in 1986. This third wheel, installed in May 2013, features higher-quality materials and is expected to last 50 years. Pumped water flows constantly over the wheel at 20 gallons per minute. When the mill is in operation, the flow is increased to 200 gallons per minute. (Author's collection.)

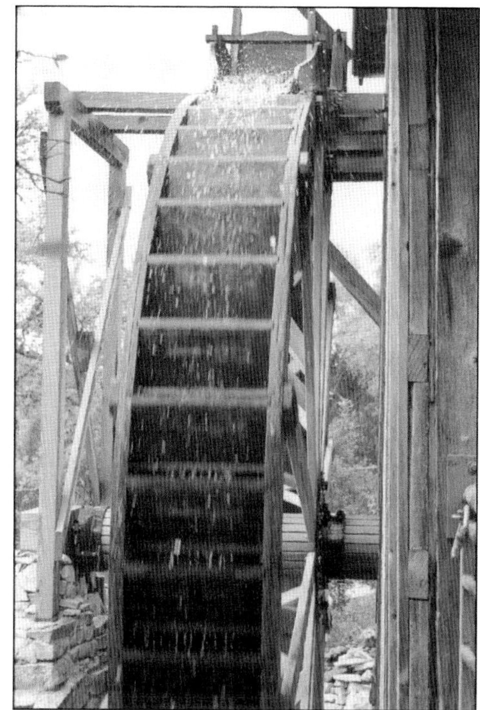

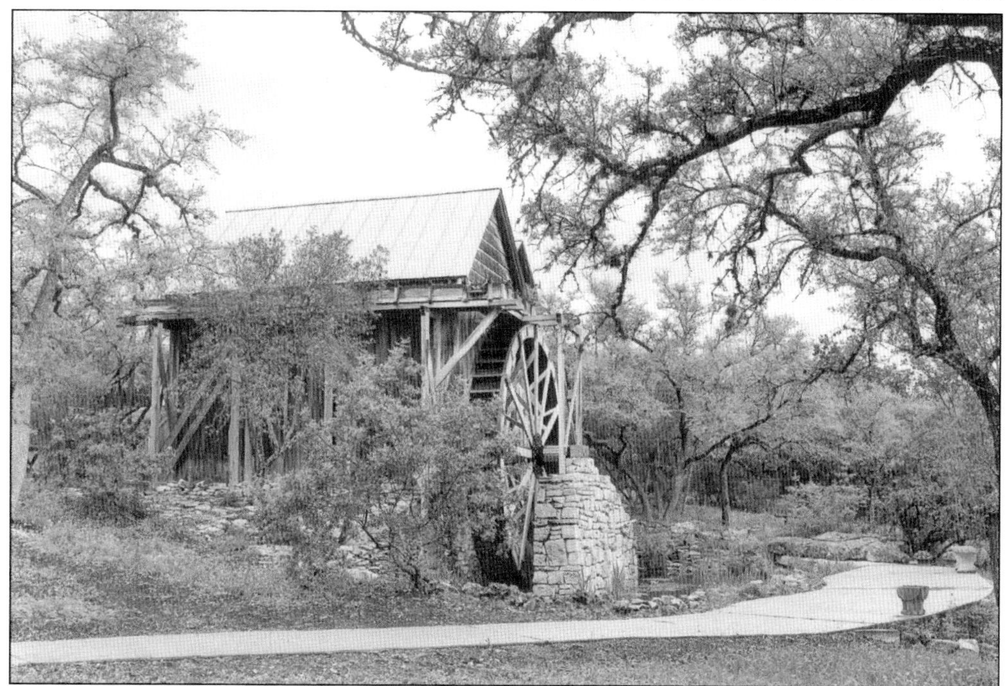

In 1965–1966, the Anderson Mill Garden Club completed construction of a full-size operational model of the mill. The mill and museum are open to the public on the fourth Sunday of each month from March through October. It is located at 13974 FM 2769 in Volente. Members of the garden club and those who assisted are to be commended for restoring this vital part of Texas milling history. (Author's collection.)

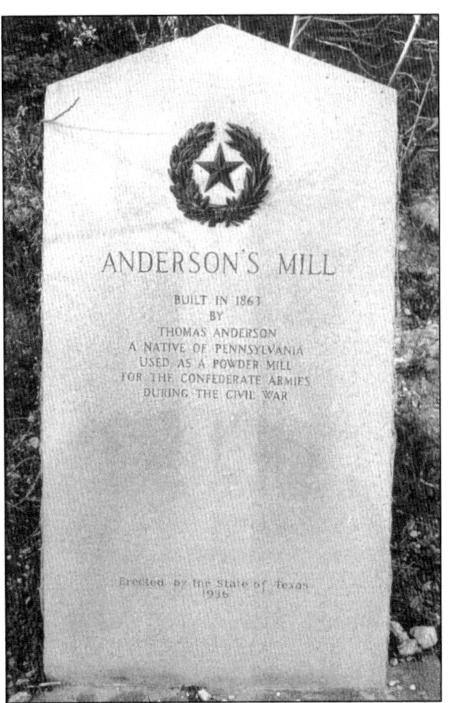

A historical marker erected in 1936 commemorates the significance of Anderson's Mill to the settlement of the Austin area. W.D. (Bill) Anderson, grandson of Thomas Anderson, once stated, "Historical markers are like certificates of merit or diplomas—impressive, but somehow they fail to reveal the drama behind the scenes—the love, the labor, the warmth and the hospitality, the successes and failures, the tears and the laughter that are a part of all human undertakings." (Author's collection.)

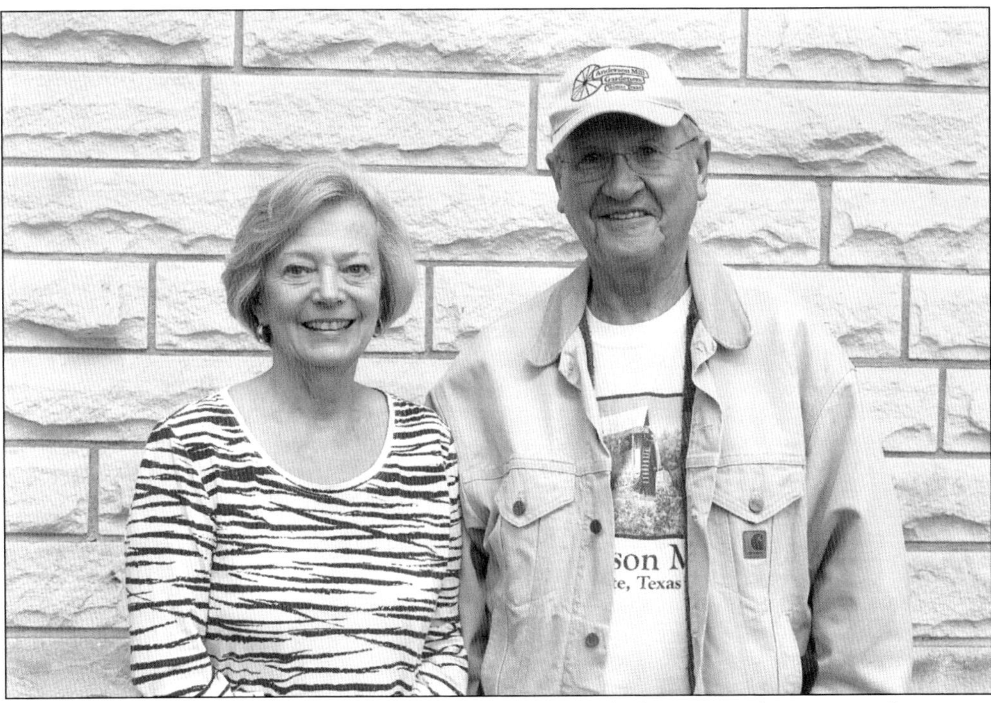

Charlynn Anderson Casey, great-great-granddaughter of Thomas Anderson, stands next to Roger Shull, chairman, Anderson Mill Memoria. Shull was at the mill to give the author and her husband a tour of the mill. Casey had driven from Temple, Texas, to attend an art class sponsored by the garden club. Four people who had not known each previously thus became acquainted at the mill. (Author's collection.)

Four

The Brazos River

The Arms of God

The Brazos River proper rises in Stonewall County, where the Double Mountain Fork joins the Salt Fork. A third upper fork, the Clear Fork, joins the main stream farther east, near Graham. From there, the river flows generally southeast and empties into the Gulf of Mexico near Freeport. From its headwaters in New Mexico to its mouth, the Brazos River has traveled about 1,280 miles, making it the longest river to flow through Texas.

There are many legends explaining how the Brazos got its name. Most involve Spanish explorers who had exhausted their supply of drinking water and, when at death's door, by an act of Providence, reached this flowing stream of fresh water where they quenched their thirst. In gratitude, they christened this unknown stream Rio de los Brazos de Dios, meaning "The River of the Arms of God."

The canyons of the upper-middle region of the Brazos were home to Apache and Comanche Indians. One of the millers featured in this chapter set up a trading post along the Brazos near the Paluxy River. Charles Barnard established a friendly, trusting relationship with the Indians of this region. One day, the Comanche Indians brought in a young Spanish-Italian woman, called Juannah, whom they had captured as a young girl. After keeping Juannah captive for three years, they brought her to the trading post, where the chief and tribal leaders exchanged her for a pile of blankets and food goods in 1846. The young girl was finally free. She later became the wife of Charles Barnard, who established Barnard's Mill at Glen Rose.

The Brazos River and its tributaries powered the gristmills presented in this chapter. Those mills were located at Eliasville, Fort Worth, Paluxy, Glen Rose, Dublin, Clifton, Norway Mills, Valley Mills, Adamsville, Georgetown, Jonah, and Salado. One of the two working mills in Texas is also included in this chapter. Brazos de Dios, known as Homestead Heritage Village near Waco, is named after this river, which runs through its property. The chapter concludes with the Chenango Sugar Mill in Brazoria County.

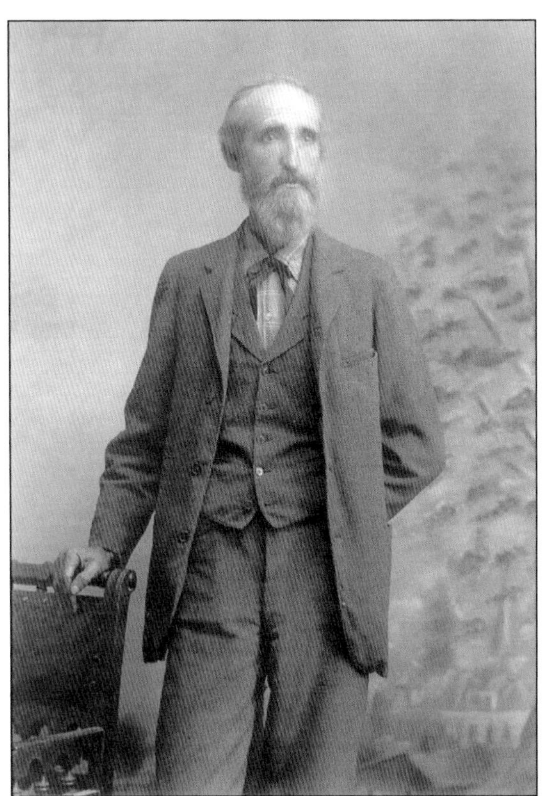

In the late 1870s, Thomas Franklin Donnell, pictured, and his brother, William Leander Donnell, along with their father, James D. Donnell, built a dam and gristmill on the Clear Fork of the Brazos River near the town of Eliasville, Young County, Texas. The Donnell brothers, both Confederate veterans, had moved from Missouri and established ranches in the area before building their mill. (Margaret Lambkin.)

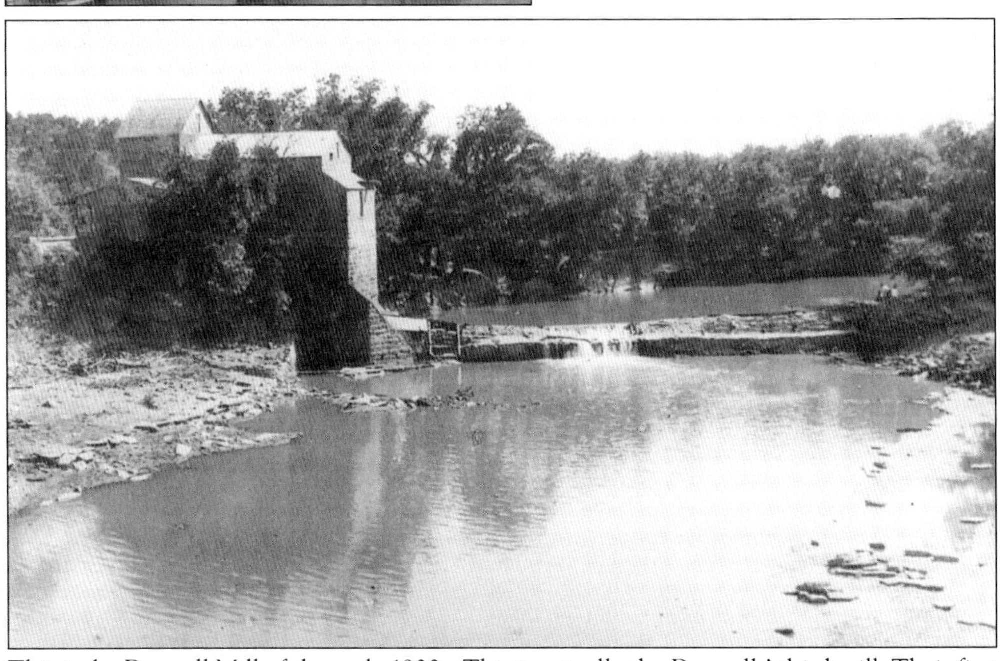

This is the Donnell Mill of the early 1900s. This is actually the Donnells' third mill. Their first and second mills were destroyed by floods. This mill, built in 1879, turned out flour and meal for a vast region and for a time was the only mill west of Fort Worth. Settlers came from a radius of 100 miles to have their wheat and corn ground. (Margaret Lambkin.)

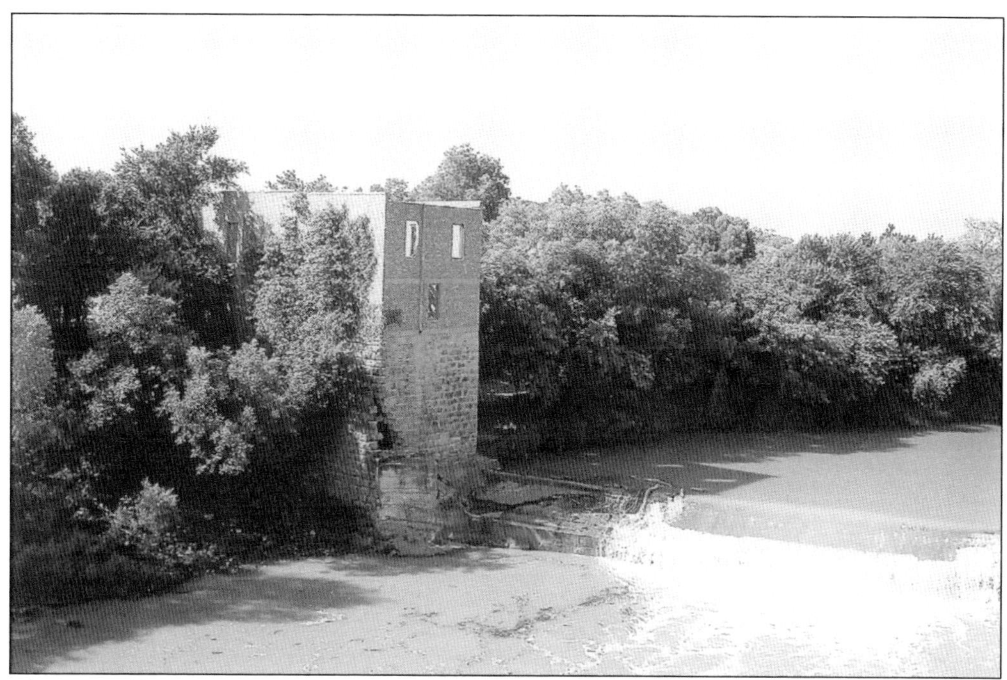

The Donnell Mill also ground meal and flour for the troops stationed at Fort Griffin, a military post on the Clear Fork in what is now Shackelford County. The mill grounds provided a gathering place for area residents, hosting such events as traveling circus shows and religious revival meetings. In 1915, Walter A. Andrews bought the mill, and he operated it until it burned down in October 1927 when it was struck by lightning. Andrews rebuilt the mill and operated it until his death in 1941. Above is the mill as it appeared in 2008. Below is the mill as it appeared in 2014. Even though the mill is nonoperational, the well-built structure still stands on the Clear Fork River in its original location. (Both, Margaret Lambkin.)

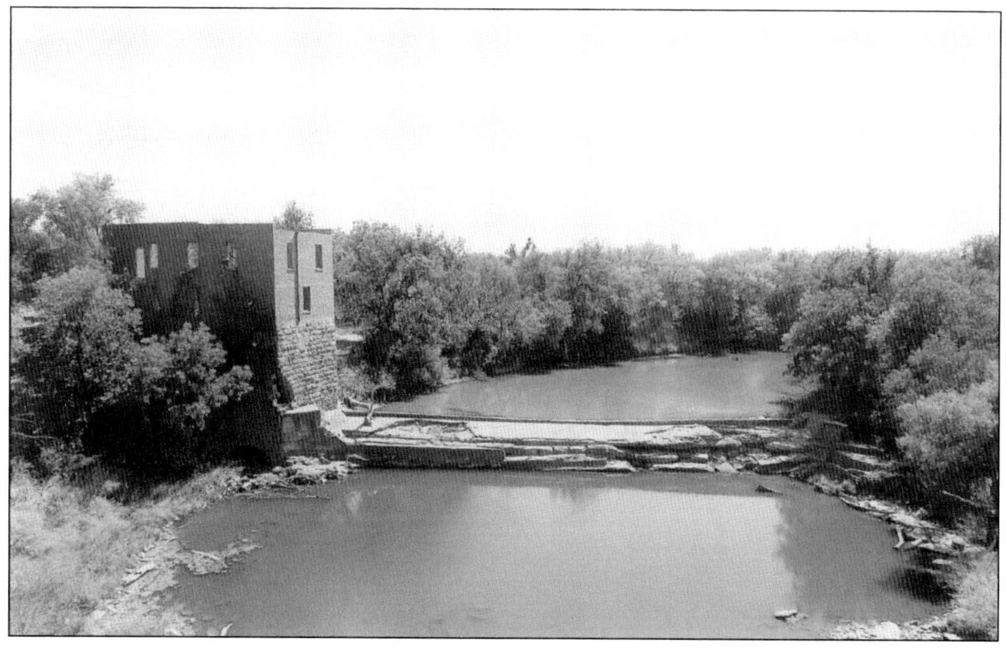

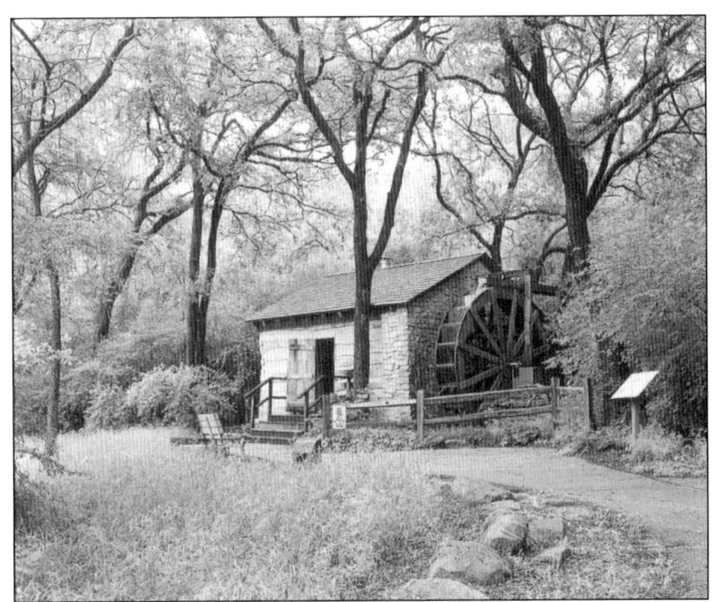

Thomas J. Shaw built his cabin in 1854 in the Spring Creek community of Parker County. When first erected, there was not another pioneer settlement west of his cabin. Shaw's nearest neighbor was a group of Tonkawa Indians camped along the Brazos River. After more than a century of use, the cabin was relocated to the Log Cabin Village in Fort Worth, where it is now being preserved as a gristmill for educational purposes. (Author's collection.)

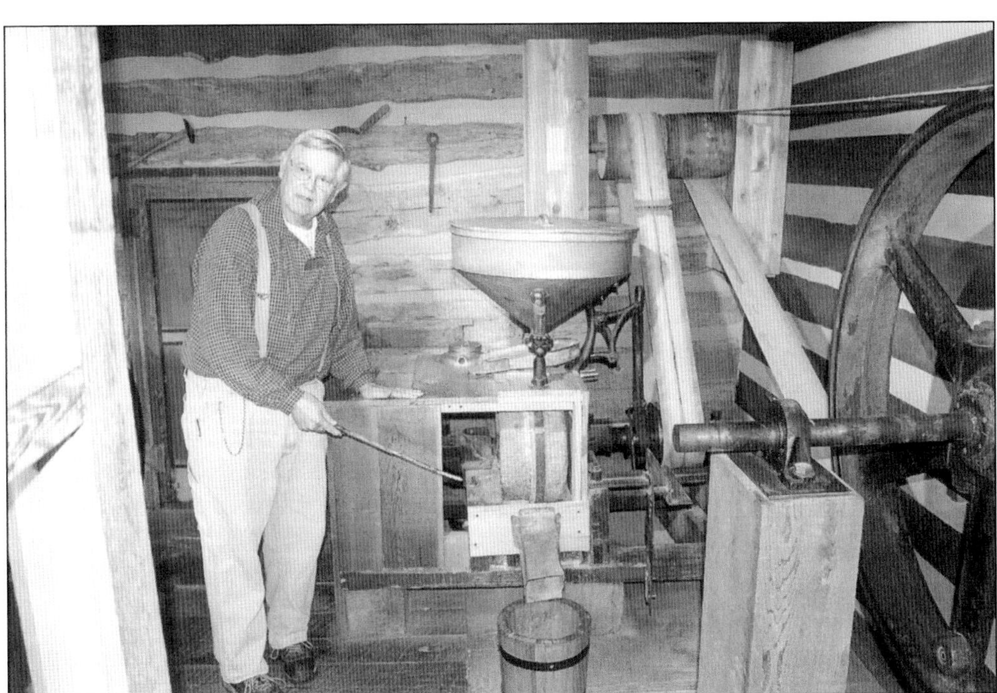

Terry, a historical interpreter and the miller at the Shaw Cabin and Gristmill, explains that the 1860s milling equipment came from a small mill near the town of Moline, Texas. The mill had been in continuous use for over 70 years before operations ceased in the 1930s. The Shaw Cabin and Gristmill is one of nine historic structures at the village where visitors can experience life as it was in the mid-1800s. (Author's collection.)

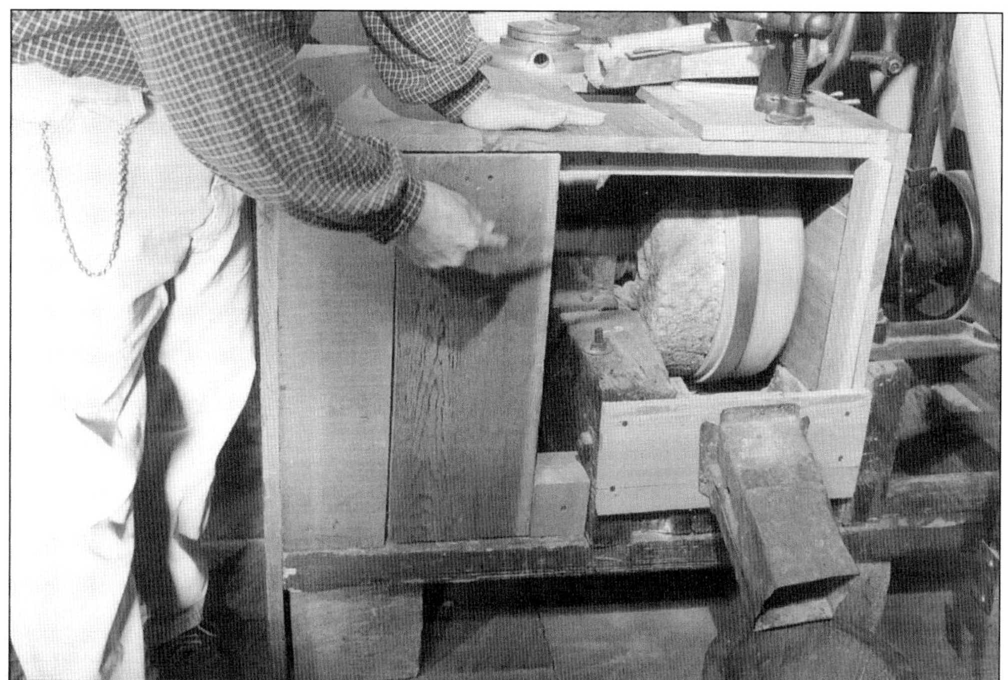

The gristmill, using the original millstones, continues to mill cornmeal just as it did in the 1860s. Within the millstone box are two vertical gray granite millstones. They weigh approximately 200 pounds each and are about 15 inches in diameter and about four to six inches thick. During the milling process, the corn is crushed and delivered through a metal spout as fresh, golden cornmeal. It is then bagged in burlap sacks for marketing. (Author's collection.)

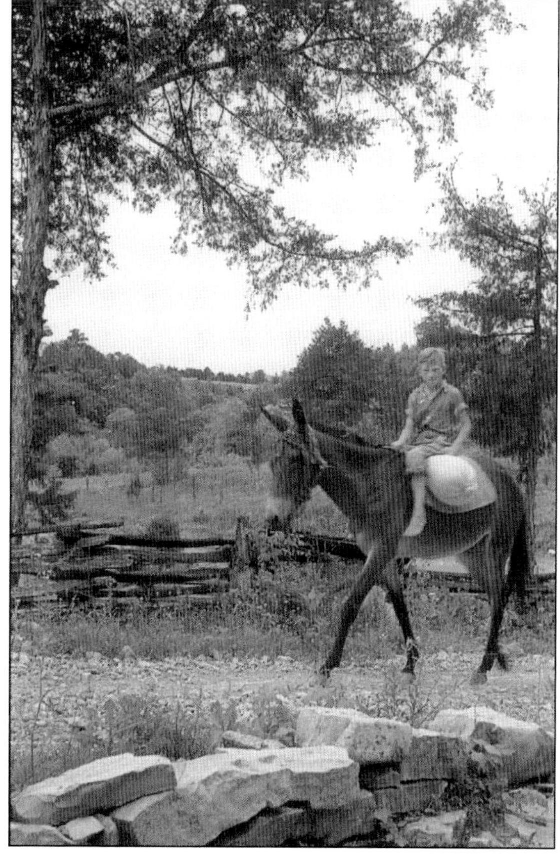

This boy is taking a sack of grain to the local mill. Youngsters delighted in going to the mill. They loved the smell of the place, seeing the white dust that covered everything, and listening to the rumble of the machinery. The biggest thrill was watching the big waterwheel with the water pouring over it. (The Portal to Texas History, texashistory.unt.edu, University of North Texas Libraries, Special Collections.)

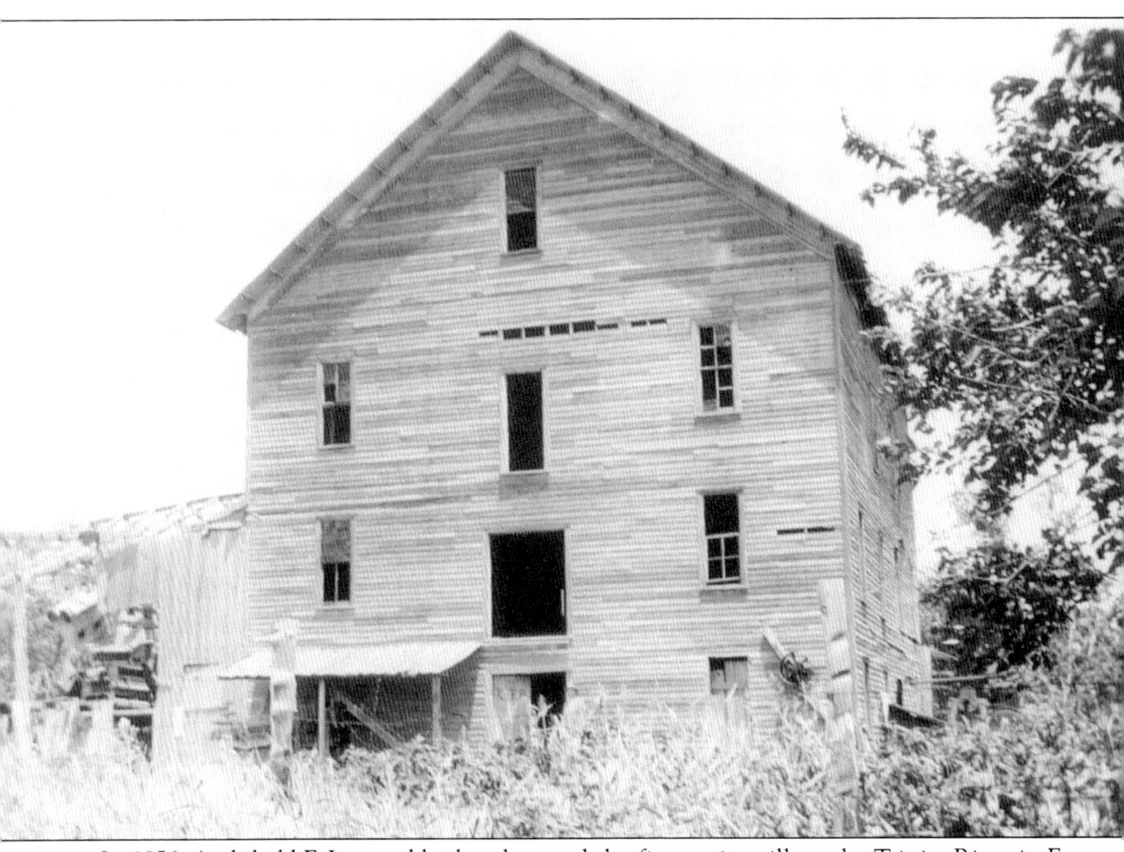

In 1856, Archibald F. Leonard built a dam and the first grain mill on the Trinity River in Fort Worth, Tarrant County. The mill became a community center and county voting place. During widespread abolition violence in 1860, the mill was burned. It was reopened by 1862 and operated during the Civil War. Owners after the war were H.B. Alverson and J.H. Wheeler. R.A. (Bob) Randol acquired the mill in 1876. The mill, a circular saw, and a cotton gin were powered by a water-driven turbine. A reminder that a mill can be a hazardous place in which to work, John Randol, the brother of the mill owner, was caught in the machinery and crushed to death in 1898. A brother-in-law, Jim Clark, fell from the top of the mill to his death. R.A. Randol died in December 1922 at the age of 73. The mill had been closed for two years prior to his death, and it never operated again. (The Genealogy, History, and Archives Unit, Fort Worth Library, Randol Mill, L-0088.)

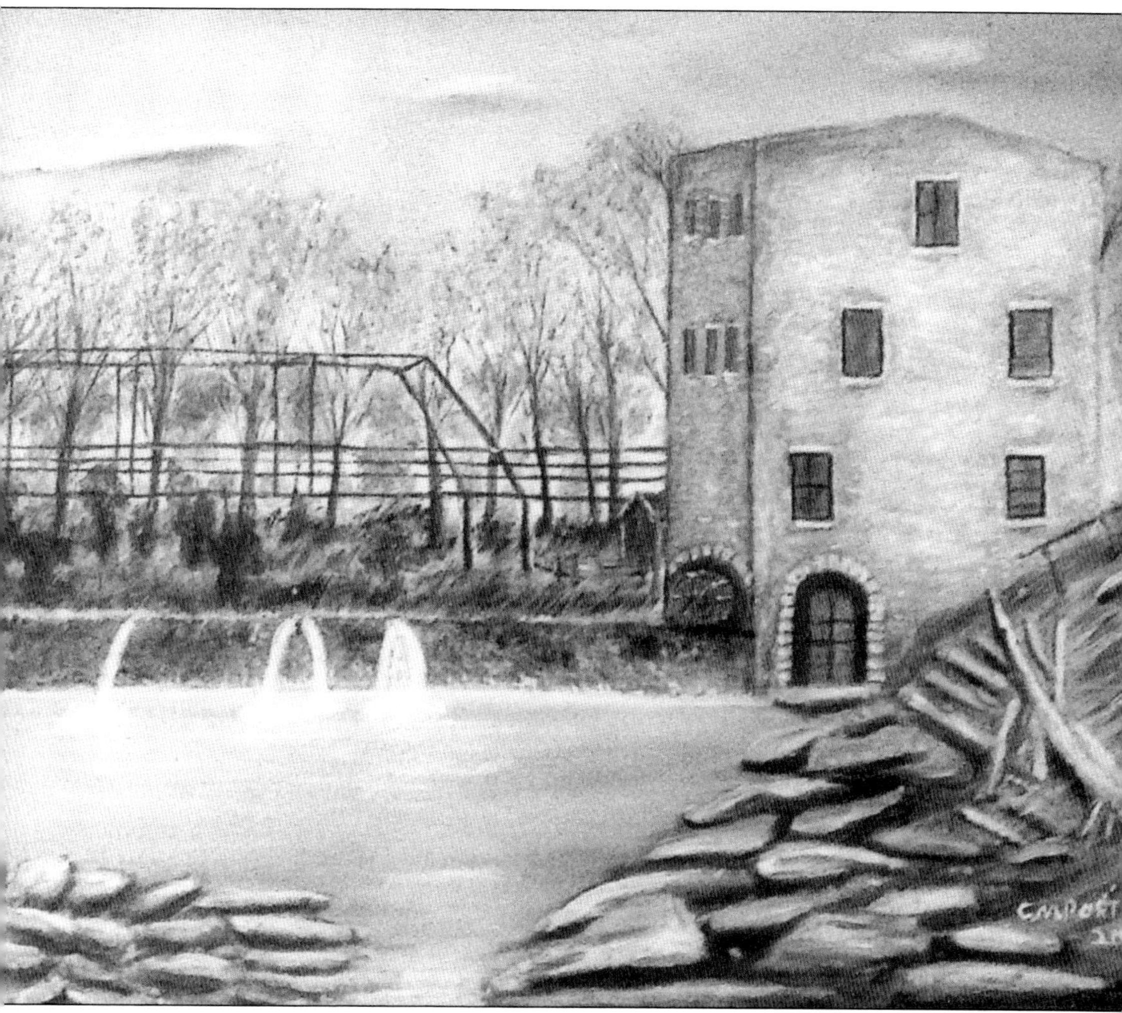
C.M. Porter, in his painting of the gristmill on the Paluxy River, has captured the essence of the mill that in days gone by met the needs of people who lived near the settlement of Paluxy, Hood County. The mill, operated by a waterwheel, was built by William Gotcher about 1860. It was built in the riverbed and was constructed of hand-hewn limestone quarried from local hillsides. Water to power the mill flowed in through the arched doorway. The waterwheel was at ground level. The second floor held the grain stored for milling, and the third floor was the residence of the miller and his family. During the mill's working days, in the 1860s to the early 1900s, farmers from miles away brought corn and wheat to be ground. In 1908, a devastating flood along the Paluxy River demolished the mill and the bridge that ran alongside it. People waited eagerly for the mill to rebuild and begin operating again, but it never did. Consequently, the settlement of Paluxy went from a bustling milling town to the small, quiet community that it is today. (Patricia White.)

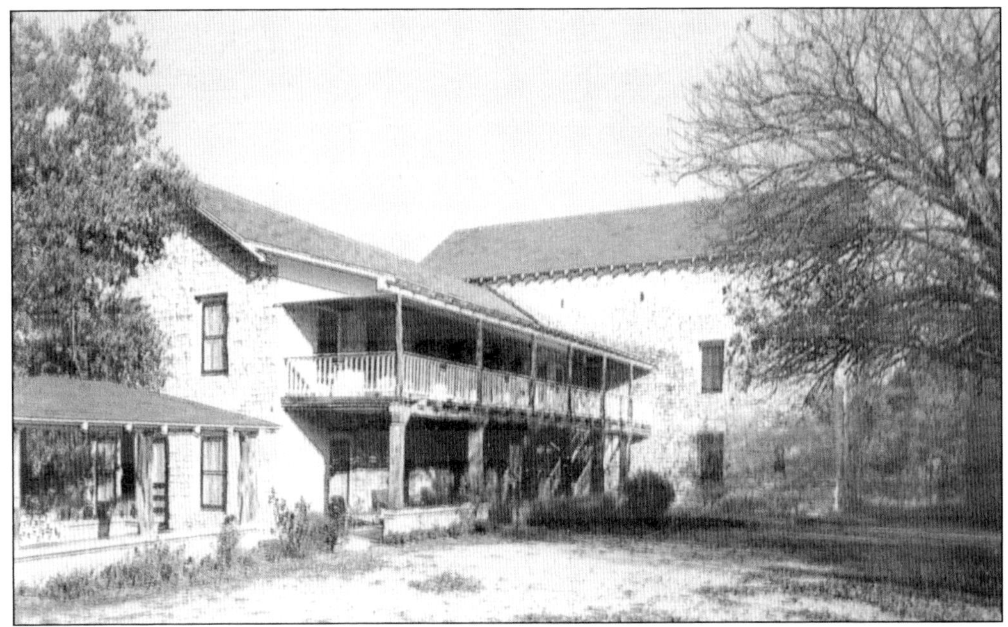

In 1860, Charles Barnard contracted for a gristmill to be built along the Paluxy River in present-day Glen Rose, Somervell County. The mill was built like a fort, with 30-inch-thick stone walls and gun ports on the third floor at 10-foot intervals. The family home, with its double porches, was built adjacent to the mill. Limestone was quarried from the nearby mountains and hauled by ox-drawn wagons to the building site. (Somervell County Museum, Glen Rose.)

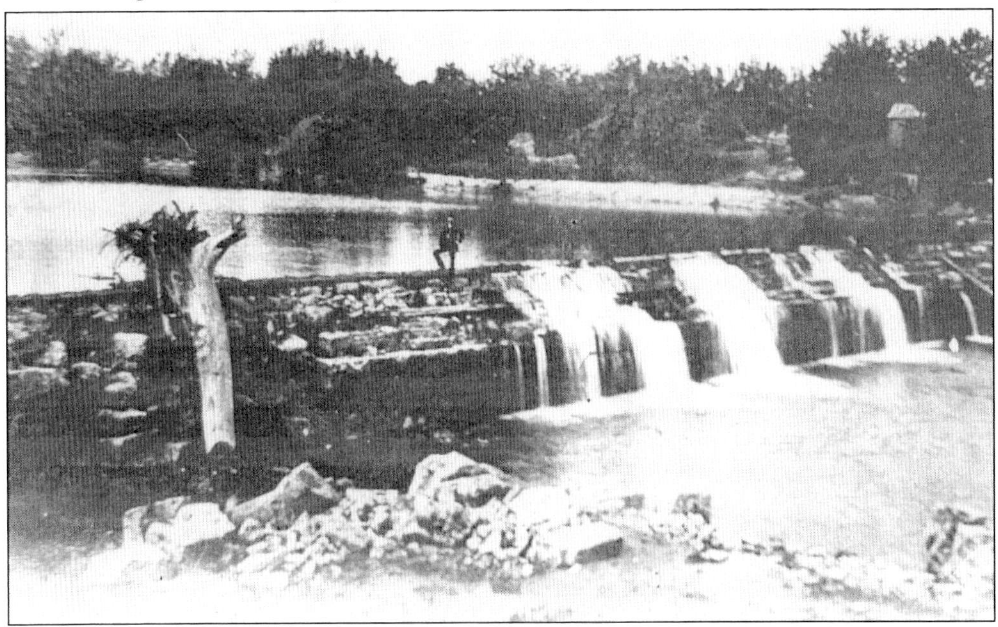

A dam was built on the Paluxy River to furnish power for Barnard's Mill. After the dam was constructed, Barnard built a millrace to power the turbojet waterwheel in the basement. An outside waterwheel never existed. After four year of construction, Barnard's Mill began operation in 1864. The mill operated on waterpower until 1890, when it was converted to wood-burning steam engines. (Somervell County Museum, Glen Rose.)

Charles E. Barnard (1823–1900) came to Texas from New York as a young man in mere search of adventure. He and his brother George set up trading posts in the Brazos River valley area and developed a good rapport with the local Indians. Barnard owned and operated the Barnard Mill and Trading Post until 1870. After the property sold, the name was changed to Glen Rose. (The Granbury Junior Woman's Club.)

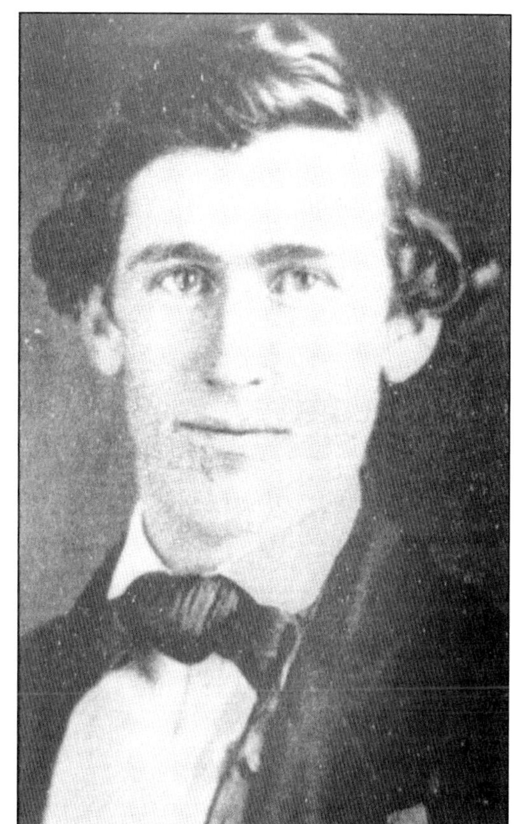

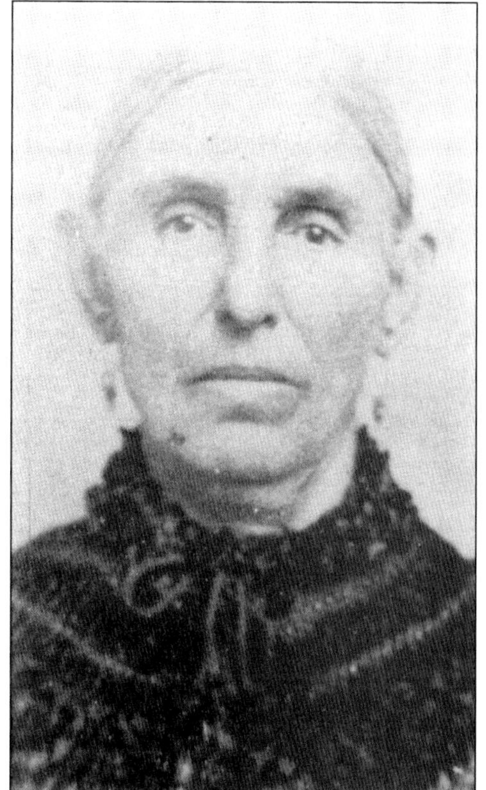

Juannah Josephine Cavasos was captured by Comanche Indians while visiting a girlfriend in Texas, across the border from Matamoros. After three years, in 1846, an Indian raiding party made its way to Barnard Trading Post at Waco Village. There the Comanches traded Juannah for supplies. Two years later, she and Charles Barnard married, and they were the first settlers in Hood County. (The Granbury Junior Woman's Club.)

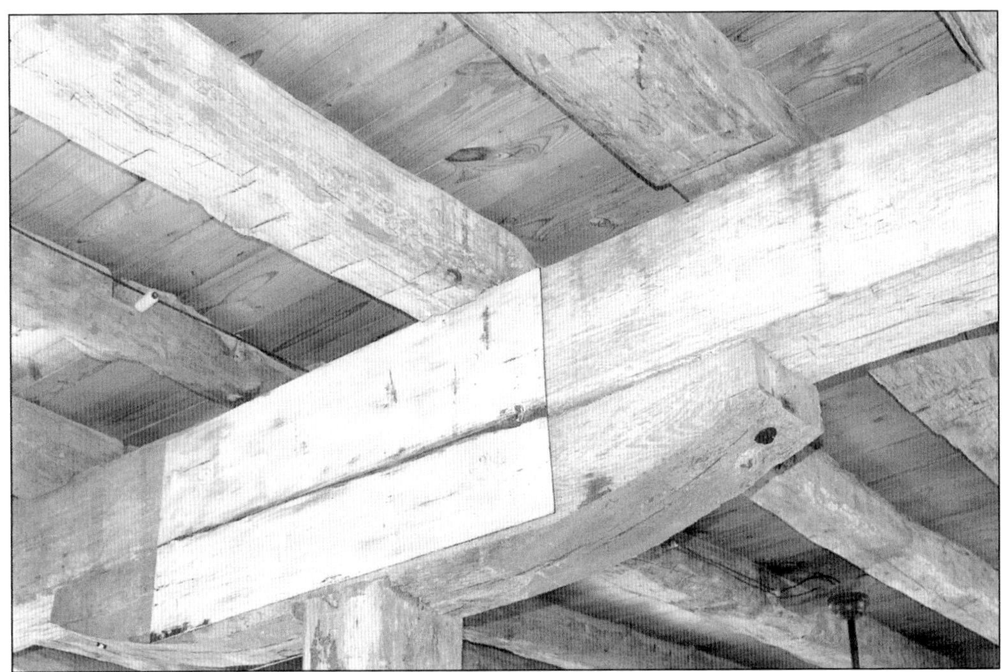

The second and third floors of the mill are supported with heavy beams. These beams were hewn with axes from the massive trunks of burr oak trees growing nearby. The floors were made of pine lumber hauled from East Texas mills. (Author's collection.)

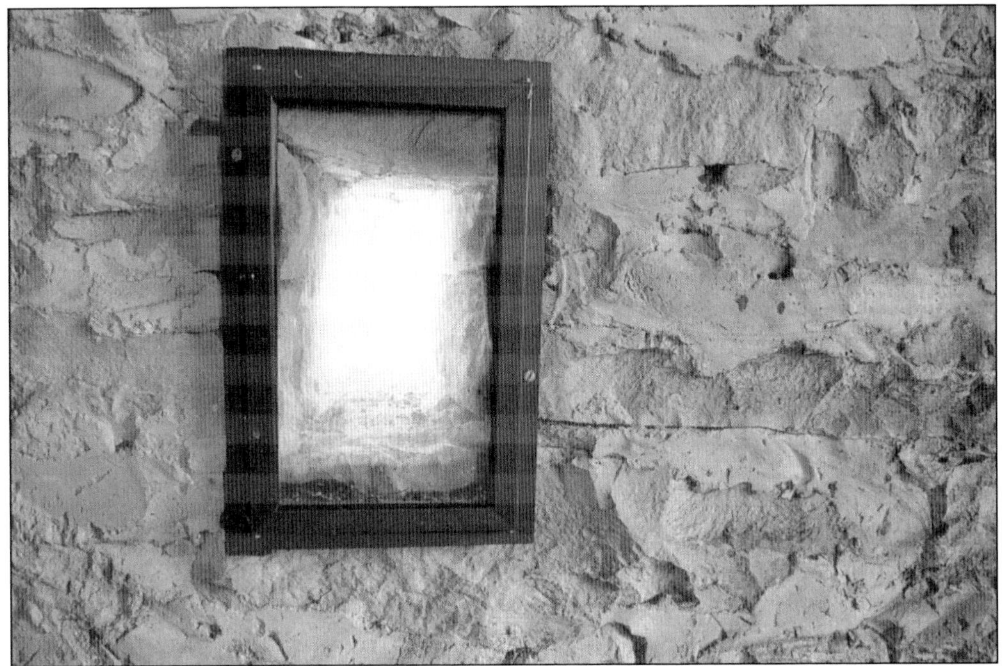

Barnard's Mill was constructed at a time when the Comanche Indians still occupied the Glen Rose area. Even though Barnard opposed actively pursuing the Indians, he built his house and mill as a stronghold to protect his family and other settlers against Indian raids. Above is a view from one of the gun ports on the third floor of the mill. (Author's collection.)

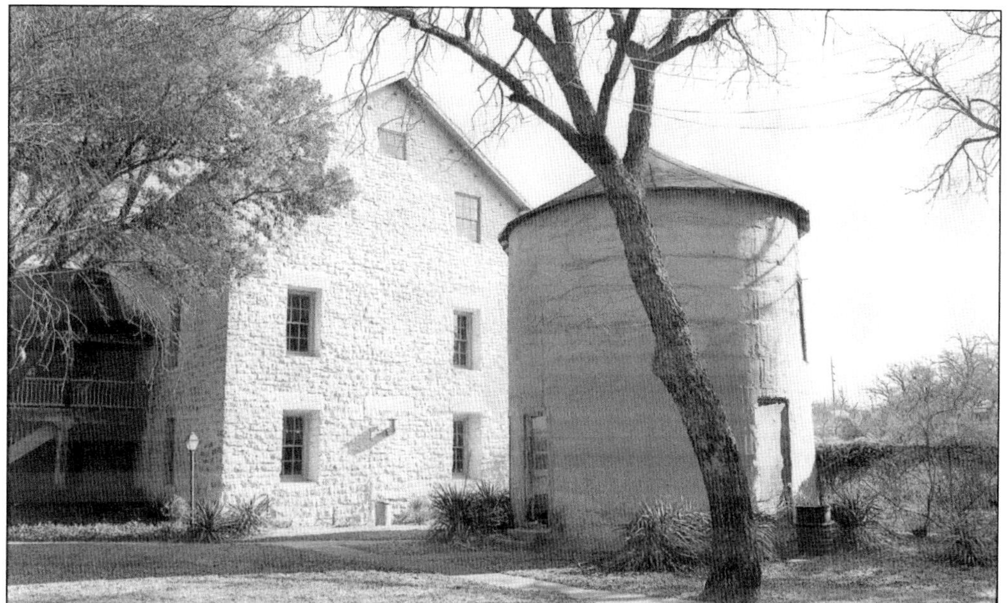

In the mid-1870s, Barnard sold the mill to Maj. Thomas Jordan of Dallas. Jordan later sold to A.J. Price, who converted the gristmill into a burr mill and cotton gin. In 1943, the Price family sold the mill property to Dr. J.J. Hanna, who operated the Clinic of Healing Waters and a medical hospital and emergency room. This view shows a silo that was added to the property. (Author's collection.)

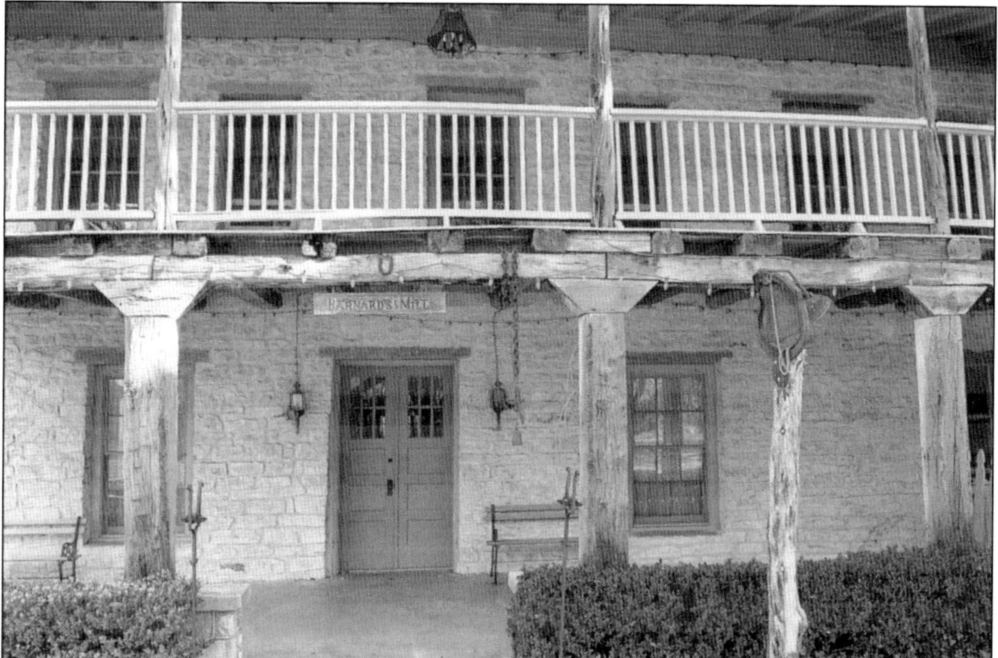

Richard H. Moore, an attorney from Fort Worth, purchased the property in 1979. Moore converted the hospital, Barnard's former home, into an art museum, and he converted the three-story mill into his private home. In 2007, Moore deeded the Barnard's Mill and art museum properties to the Somervell History Foundation. Barnard's Mill was entered in the National Register of Historic Places in 1982. This is a view of the entrance into the art museum. (Author's collection.)

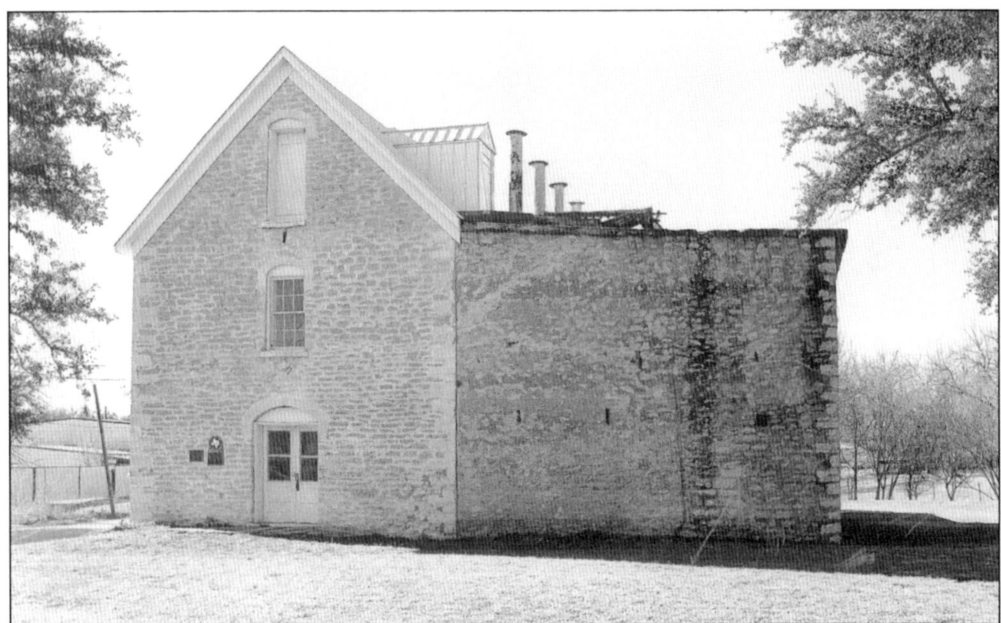

The William T. Miller Gristmill in Dublin, Erath County, was built in 1882 by stonemasons Joe E. Bishop, Rocky Davis, and Frank Hamilton. Steam power was used to grind grain until an oil engine was installed in 1918. The mill was converted to feed production after Will M. Wright and his son-in-law, Ted Robbins, purchased it from the Millers in 1926. (Author's collection.)

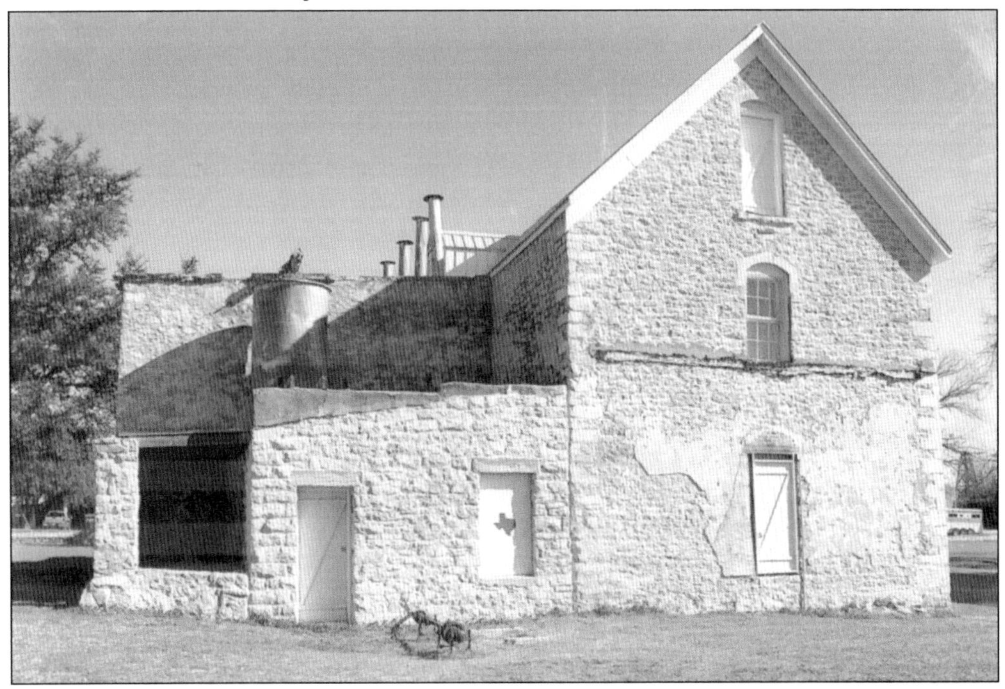

The shaft that can be seen in this back view of the mill was turned by belts that were attached to the engine that provided the power to operate the mill. The gristmill is now the oldest stone building in Dublin. The Robbins family gave the mill to the Dublin Historical Society in 1974 for use as a museum. The mill received a Texas Historical Medallion in 1974. (Author's collection.)

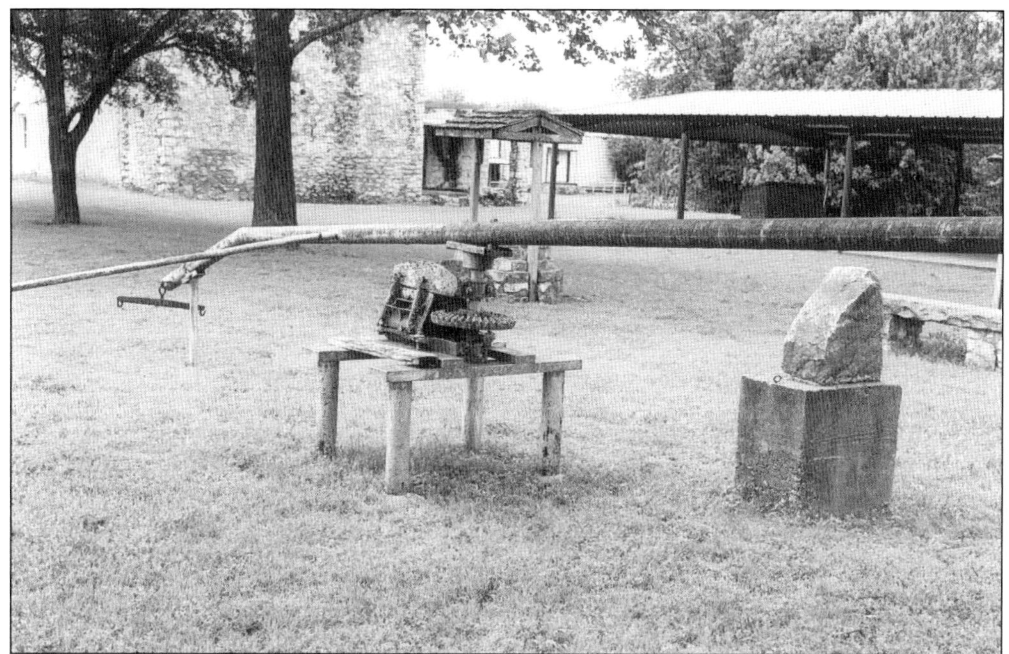

These photographs show a sorghum mill next to the Miller Gristmill. After sorghum cane is harvested, it is hand-fed into the mill a few canes at a time. Usually the mill is powered by a mule or a horse that walks in a circle. This turns the gears that operate the mill. Below is a close-up view of the mechanism of the sorghum mill. A person sits in front of the rollers and inserts the cane between them. The rollers crush the stalks, which squeezes the juice out of the cane into a container. The juice is then cooked into syrup. Sorghum syrup is usually mild and can be eaten alone, but some people prefer to pour it over hot biscuits. (Both, author's collection.)

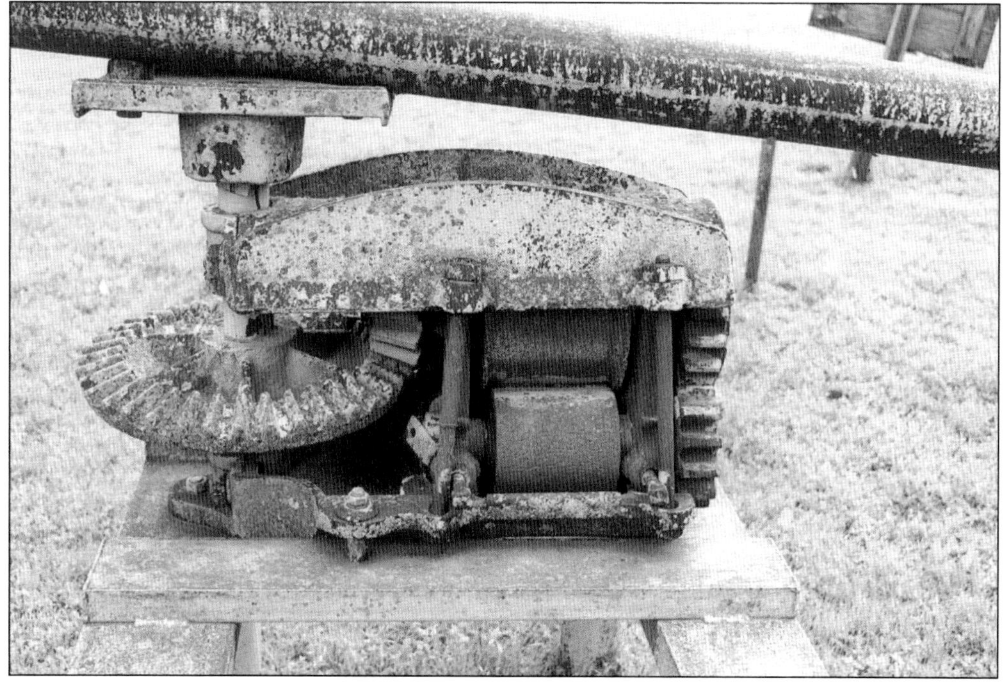

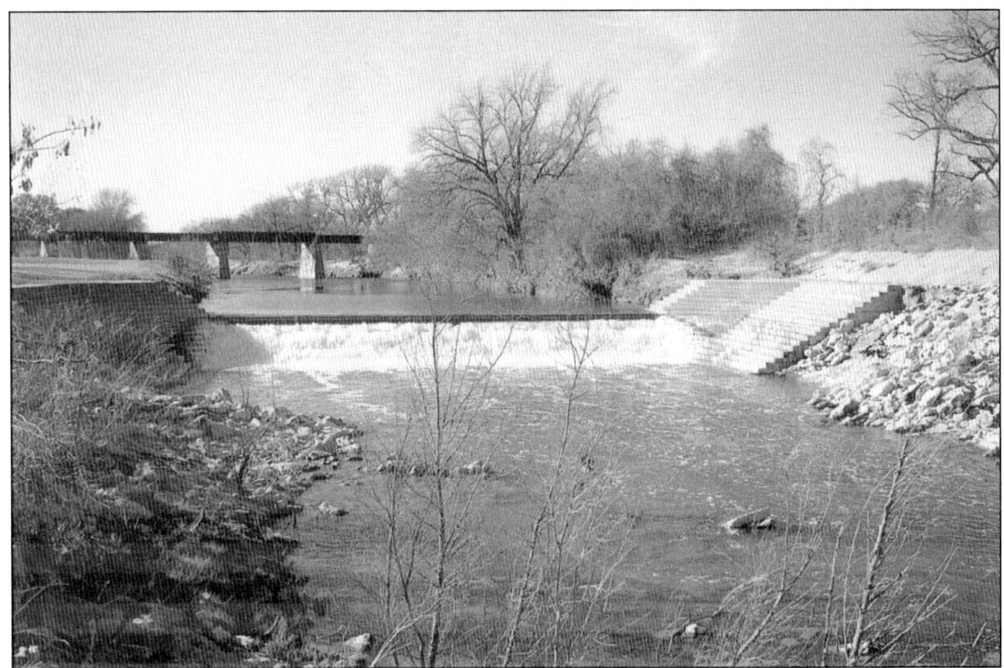

The city of Clifton had its beginning on the Bosque (BOSS-Kee) River in 1852. The Bosque River runs through several Texas counties and is fed by four primary branches. Even though the river has a rustic beauty, to the people of pioneer days, the river meant survival rather than beauty. The river offered water, wildlife, and timber for building homes. (Author's collection.)

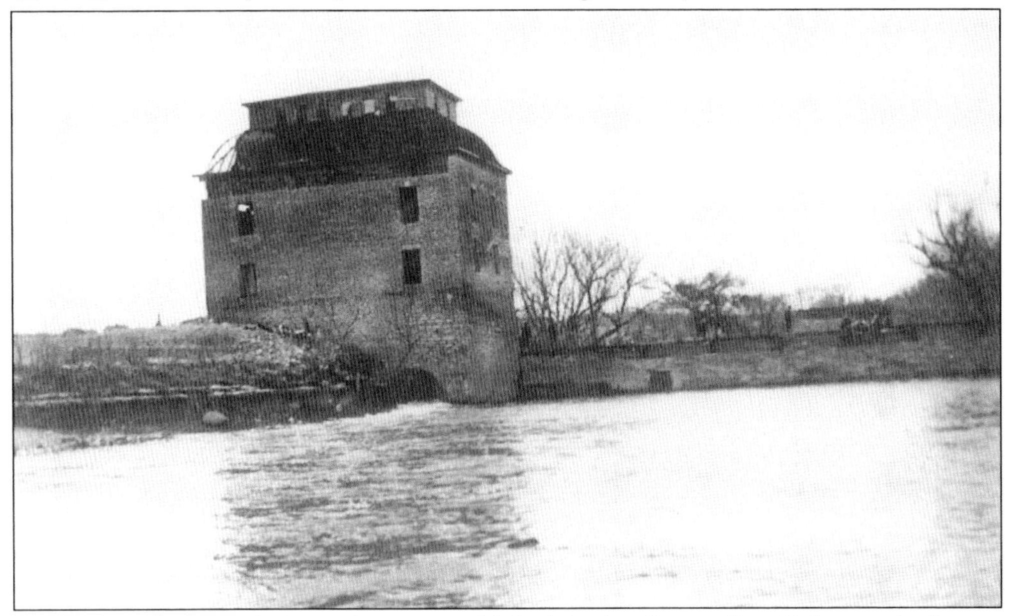

After the Civil War, Joel Martin Stinnett established the first flour mill at Clifton on the Bosque River. "The Old Mill" was built first of logs and then of limestone. It provided flour and meal to area residents and later became Clifton's first electric plant. The mill was torn down in the early 1900s but remains a beloved symbol of Clifton's past. (The Bosque County Collection, Bosque County Historical Commission, Meridian.)

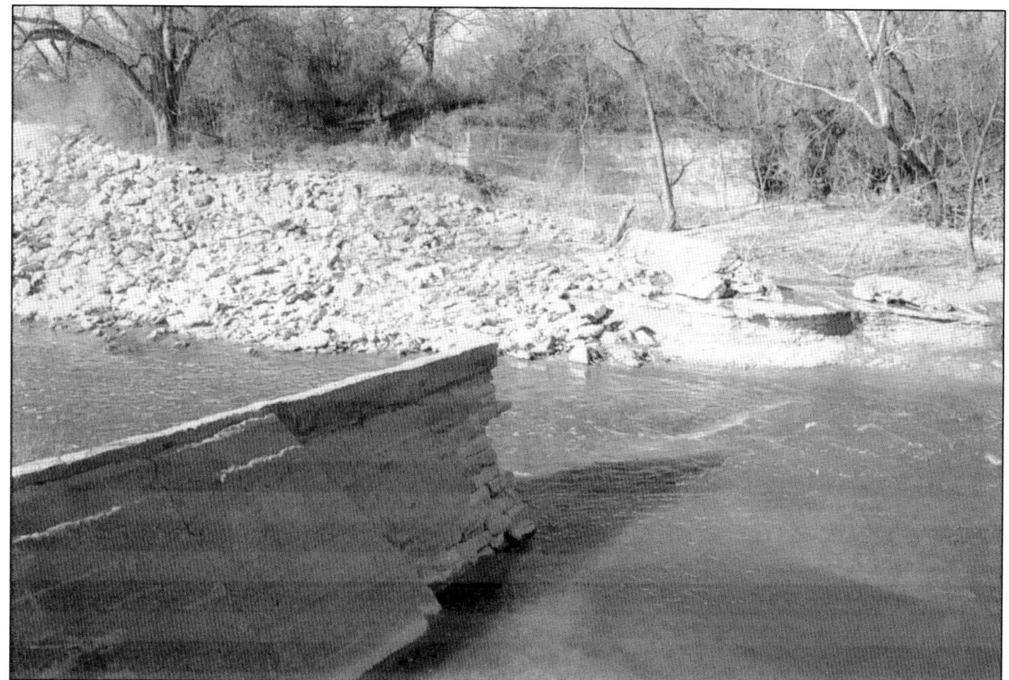

This a view of a part of the old Bosque River dam at Clifton. The Old Mill was established near this dam. The first settlers of Clifton built a wooden footbridge across the river to join pioneers on the east and west sides of the river. After a while, with the building of the railroad, the town gradually migrated to the west side of the river. (Author's collection.)

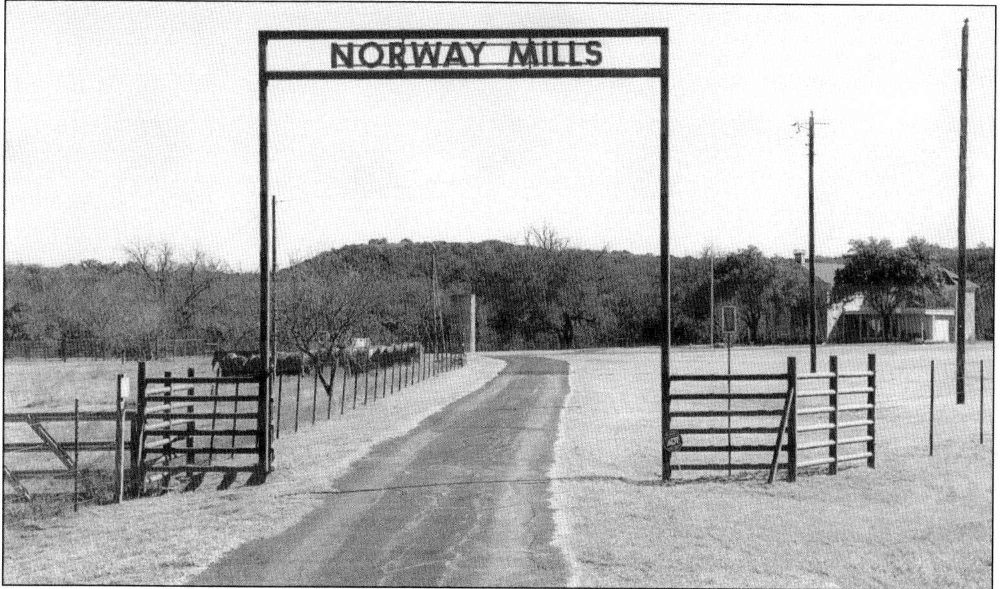

Norway Mills, near Clifton, was a community located in the Neils Creek valley. The community developed after the Civil War and flourished from 1870 to 1890. The town consisted of a schoolhouse that doubled as a church, two blacksmith shops, a general merchandise store, a drugstore, and the gallery of several photographers. Norway Mills was built and operated by partners A.Y. Reeder and Ole and Andrew Canuteson. (Author's collection.)

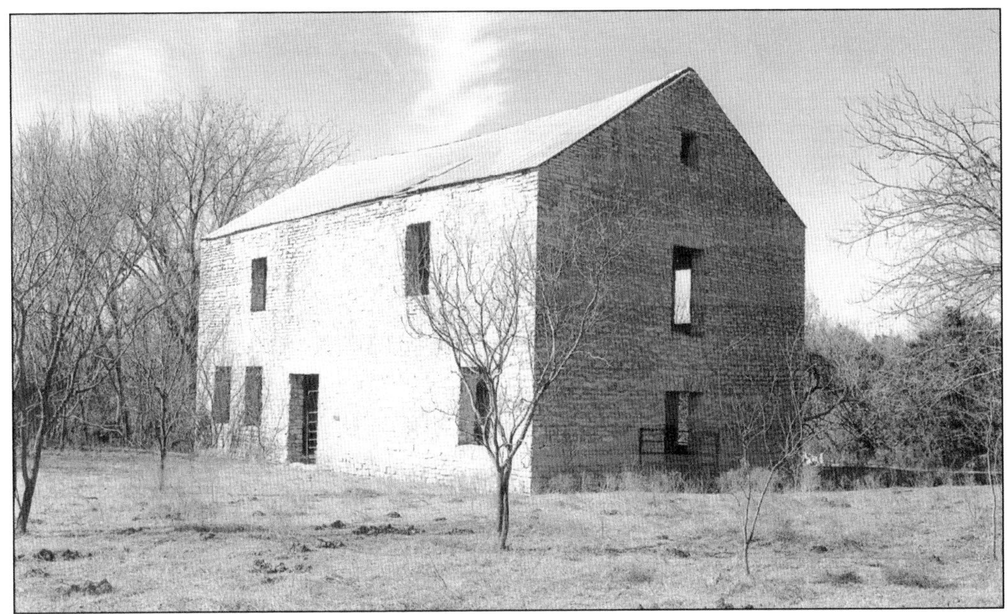

Norway Mills is a two-story stone structure. Stone for the mill and a large two-story residence was quarried in the nearby hills. The mill was steam operated. The water was drawn from a nearby branch, and the tree-covered hills supplied plenty of wood. The mill was listed in the National Register of Historic Places in 1983. The mill is pictured in January 2016. (Author's collection.)

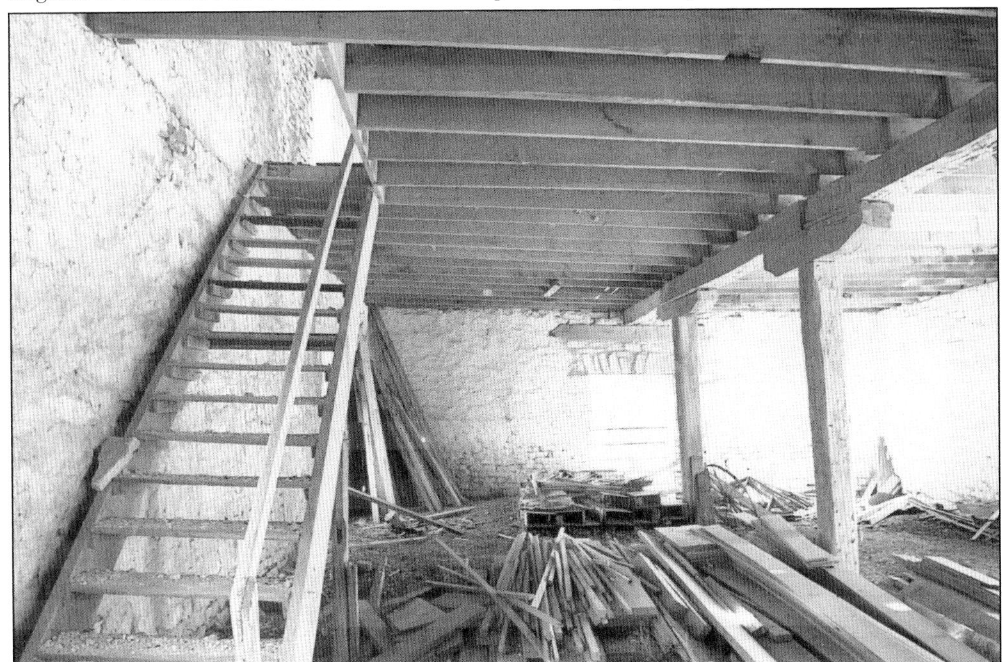

This view shows the stair steps leading to the second floor of the mill. The machinery has been removed, but records show that the mill used the burr system of grinding with two large stones. The lower stone remained stationary, while the upper stone was revolved by steam power. As the top stone revolved, the grain was crushed and ground and worked out around the edges. (Author's collection.)

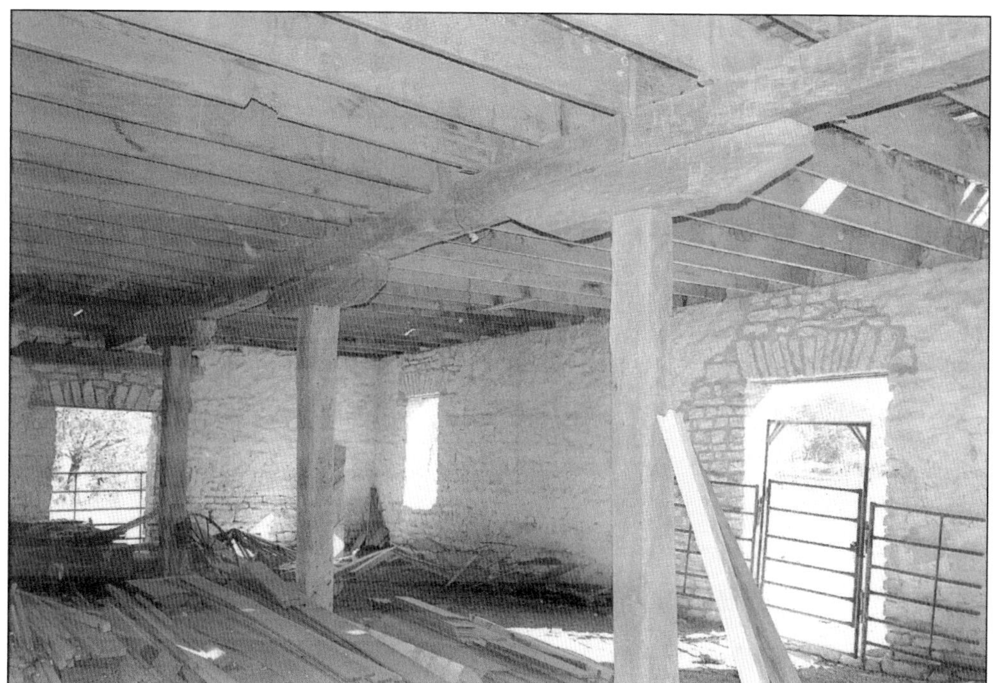

The post-and-beam inner structure of Norway Mills was reinforced with a shorter beam on the post tops to prevent sagging of the longitudinal beams. After standing for 145 years, the wood part of the structure is still in excellent condition. The metal gates were placed to prevent the ranch livestock from entering the building. (Author's collection.)

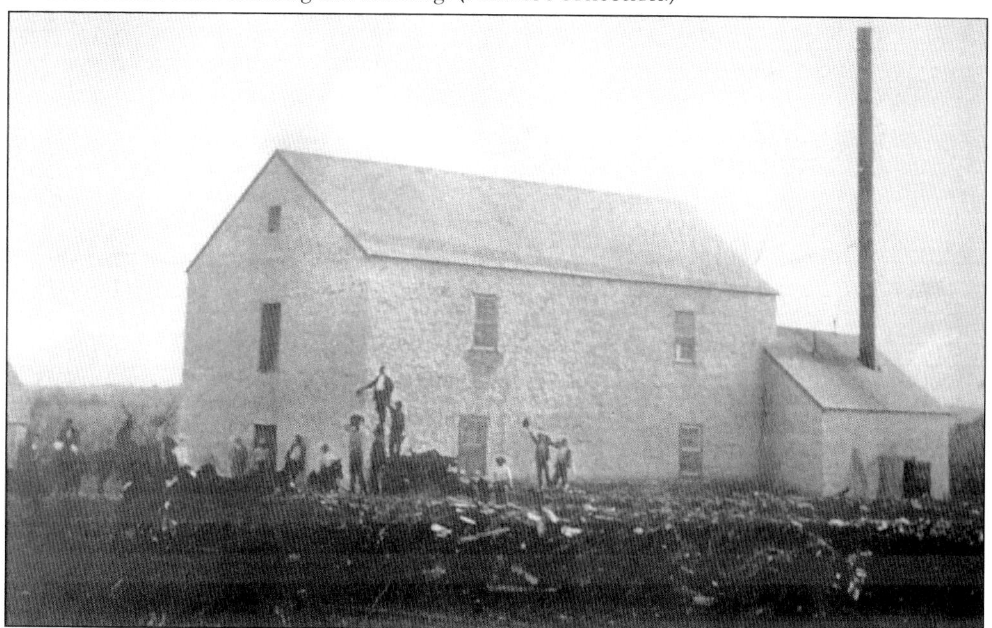

The Norway Mills cotton gin and flour mill flourished for almost 20 years before the dwindling water supply forced its closure. With the closing of the mill, Norway Mills gradually closed as a town and community center. The present owner has plans to restore the mill as a tribute to the Norwegian Texans who settled the Bosque River valley area. (Bosque Museum, Clifton.)

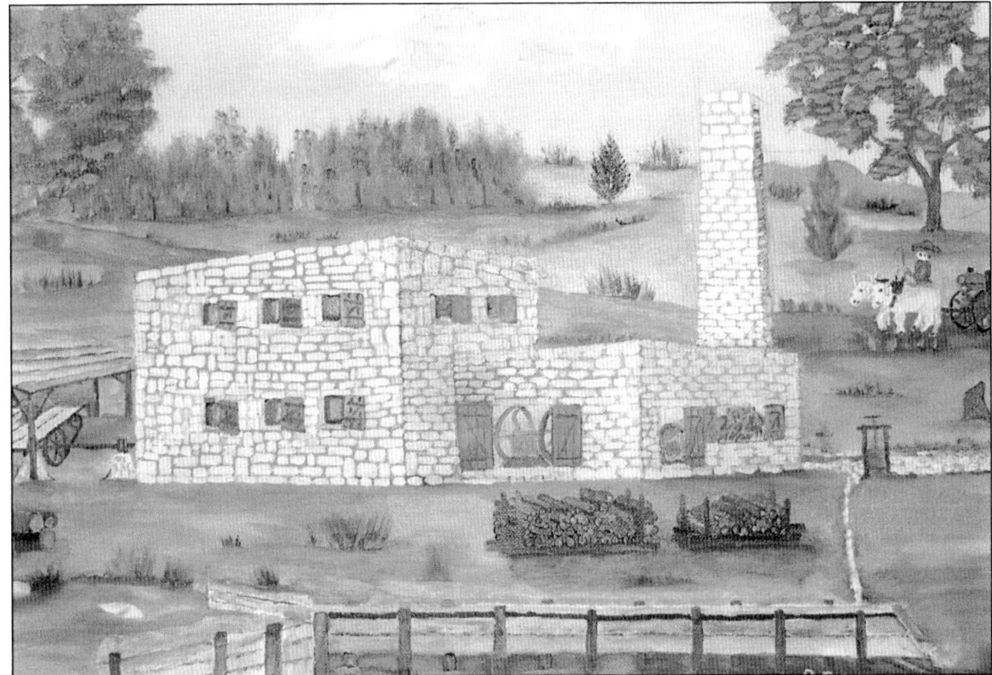

Valley Mills is on the Bosque River 11 miles south of Clifton. It was named for a flour mill established on the banks of the river in 1867 by Dr. E.P. Booth and Asbury Stegall. This painting of the mill is by R. Bowen. Based on an old photograph and written descriptions of the mill, it is displayed in the Valley Mills Public Library. (Author's collection.)

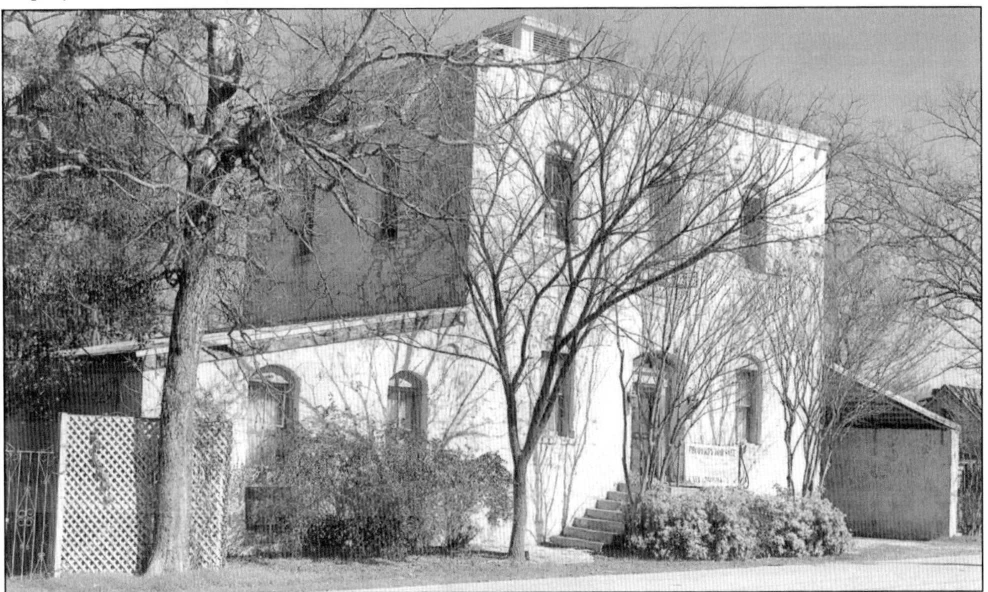

The Valley Mills gristmill ceased operation in 1917, and by 1926, all the machinery had been removed. A new owner converted it into an apartment building. In 1999, a native of Dallas bought the old building and changed it from an apartment house to her private home. During the renovation, she was careful to preserve the character of the old mill that had given the town of Valley Mills its name. (Author's collection.)

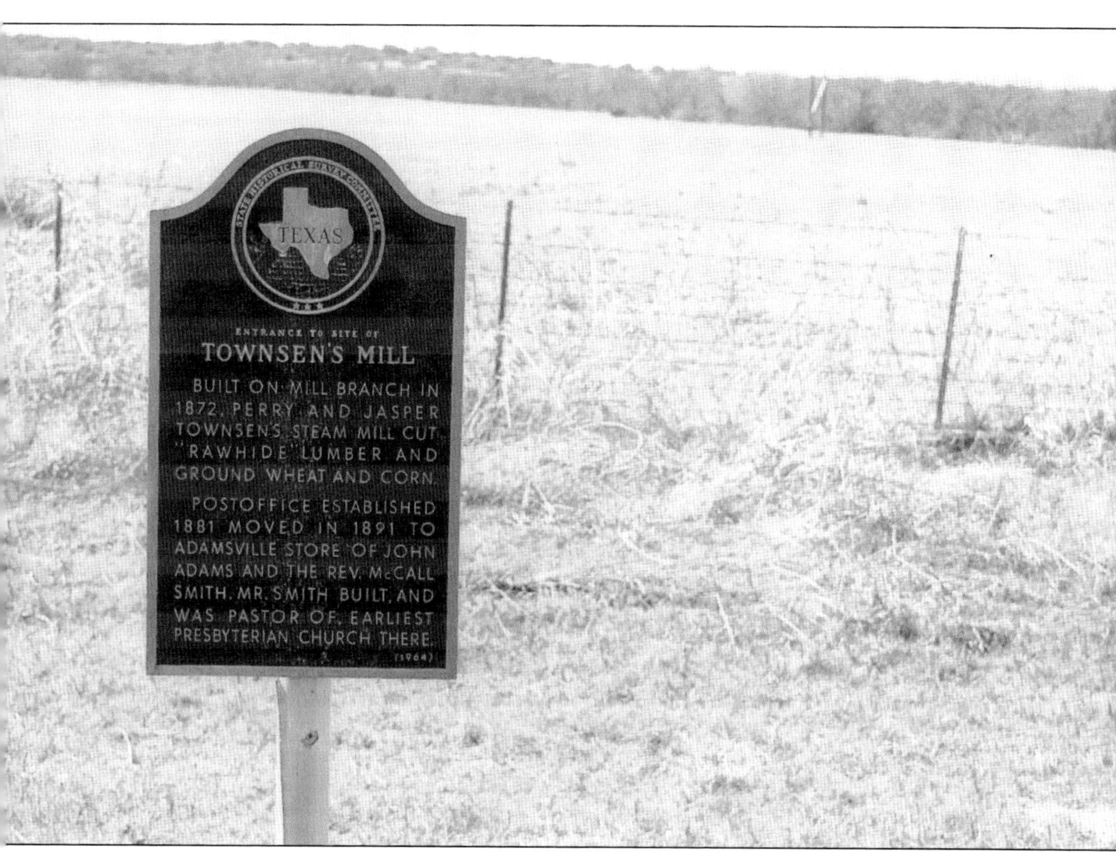

O.H. Perry Townsen came to Texas in 1854, followed by his nephew L. Jasper Townsen in 1855. In 1868, the two men bought and moved onto a small tract of land 16 miles north of Lampasas, Lampasas County. Four years later, they built a steam-powered corn mill. A sawmill and a cotton gin were added later. The mill was about a mile and a half southeast of Adamsville on a stream now known as Mill Creek that flows into the Lampasas River. This historical plaque marks the entrance to the site of the Townsen Mill. (Author's collection.)

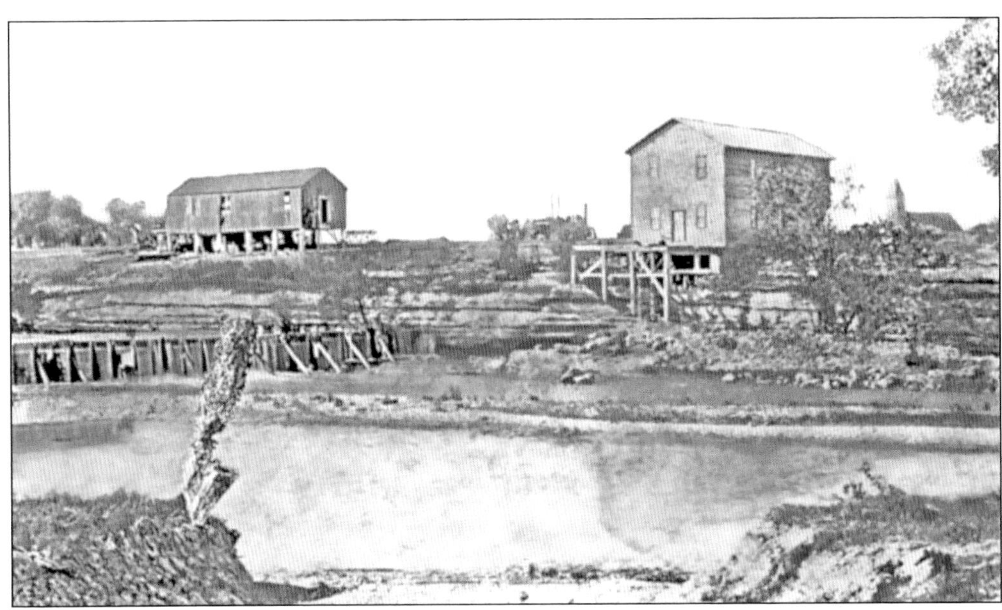

In 1857, Joseph Mileham and James Warnock built a gristmill on the San Gabriel River in Water Valley, now known as Jonah, Williamson County. The mill, known as Eureka Mills, was very productive, with slaves delivering flour by wagon as far away as Houston. The men who built the mill dam posed for the photograph below while standing on a platform over the water. The mill and the town suffered damage during the flood of 1921. The Christian Church, located west of Mileham Creek, washed across the creek to the east side. To avoid moving it back to its original site, the enterprising church members bought the land where the church building was sitting and left it at its new location. (Above, the Williamson Museum, 2001.001.9049; below, the Williamson Museum, 2001.001.9048.)

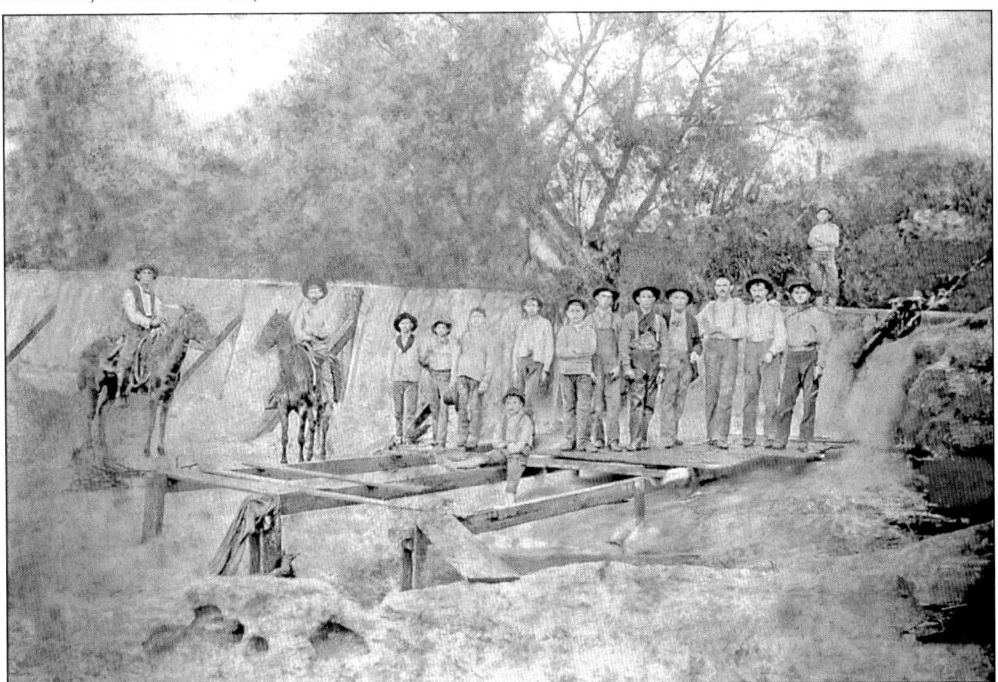

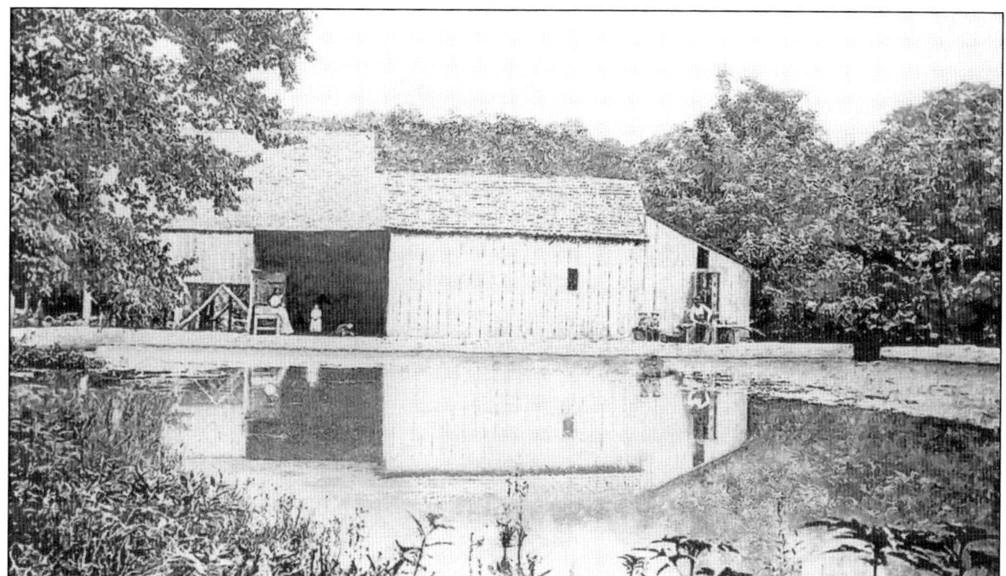

John Berry's spring-driven gristmill, built in 1848 on Berry's Creek northwest of Georgetown, was the first mill built in Williamson County. The mill was rebuilt several times over the years and was later known as Gann's (or Gan's) Mill. The mill was washed away in the 1921 flood. John and Hannah Berry and seven of their children are shown in this undated photograph. (The Williamson Museum, 2001.001.1149.)

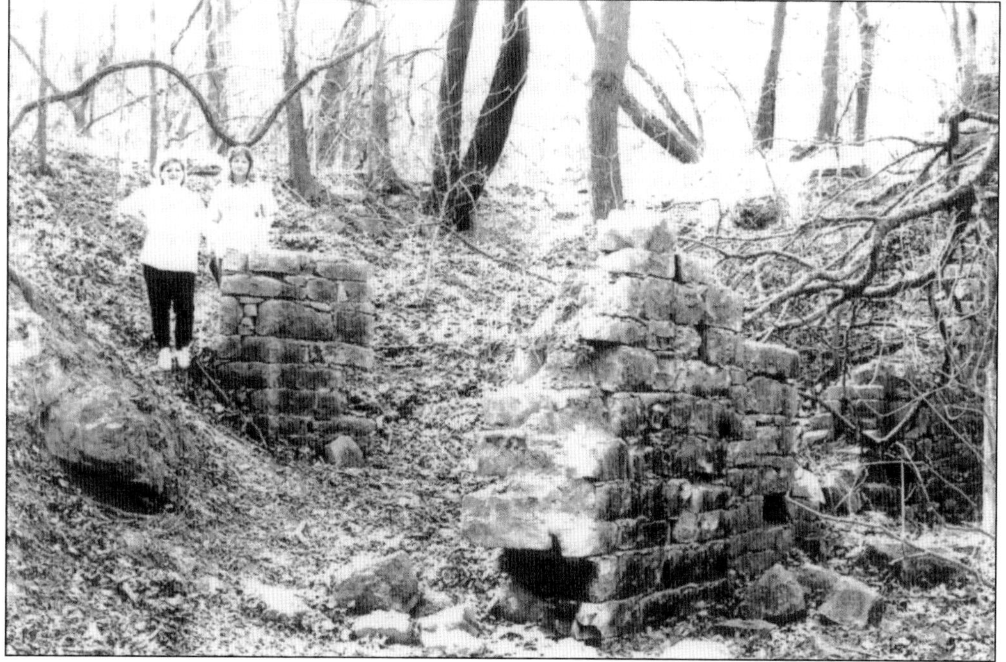

In 1848, Robert Childers and Thomas Walden built a small corn mill on the south side of the Lampasas River near present day Shanklin Road. This was the first watermill in the territory that later became Bell County. When Gordon Shanklin acquired the site in 1856, he rebuilt and modernized the mill, and it became known as Shanklin Mill. Ginny Guinn Parsons (left) and Nancy Graff Kelsey stand atop the ruins. (Kelsey Collection.)

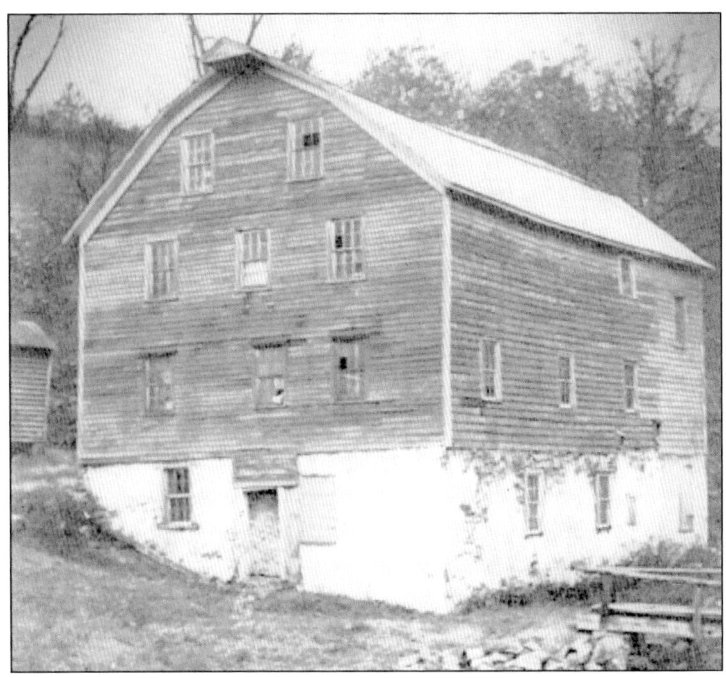

The John Teeter Mill was built around 1760 in northwestern New Jersey. At one time, it was owned by John and Gershom Mott. Soon after their purchase, the American Revolution began and American troops under George Washington camped at nearby Morristown. During the winter of 1780–1781, when the soldiers were reduced to boiling their leather shoes for food, the Mott brothers helped relieve their suffering by providing them with flour. (Ben Owen.)

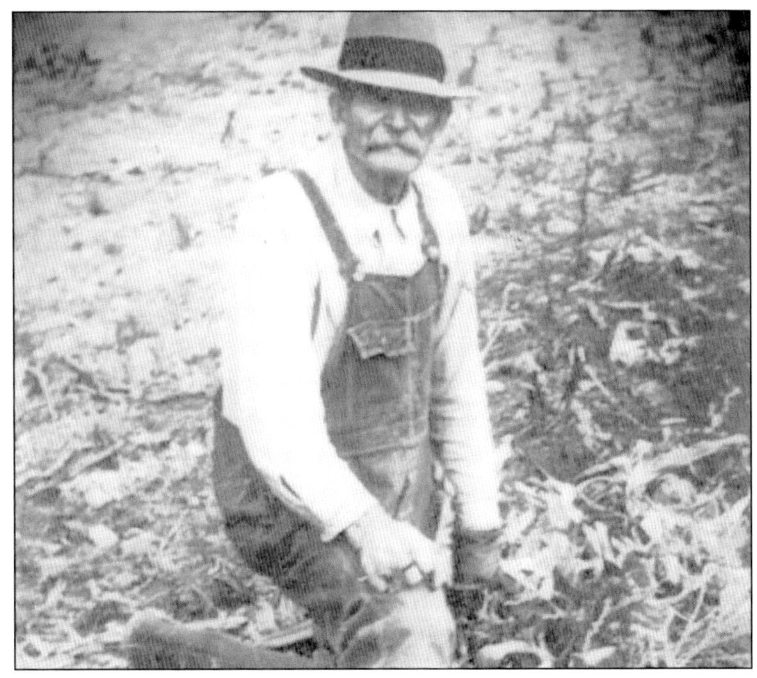

When John Teeter acquired the mill in 1814, the mill became the business and social center of a village that became known as Teetertown. The Teeter family operated the mill until 1881. Philip Sliker (pictured) purchased the mill in 1908. Ten years later, Sliker retired and closed the mill. After it last ground grain in 1918, the mill went through a series of owners before being abandoned. (Ben Owen.)

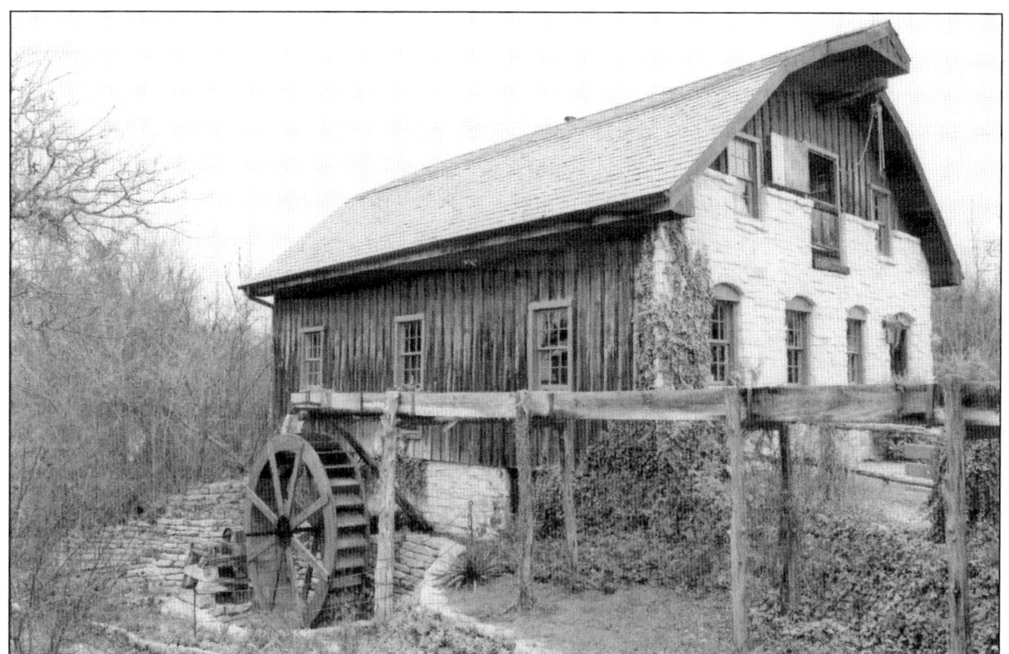

In 2001, the John Teeter Mill found a new home at Homestead Heritage Village near Waco, McLennan County. When the craftsmen of Homestead Village learned that the mill was about to be torn down, they purchased it to restore it. In order to rebuild the mill as accurately as possible, the craftsmen carefully documented and dismantled it, then brought it to Texas and restored it as a working mill. (Author's collection.)

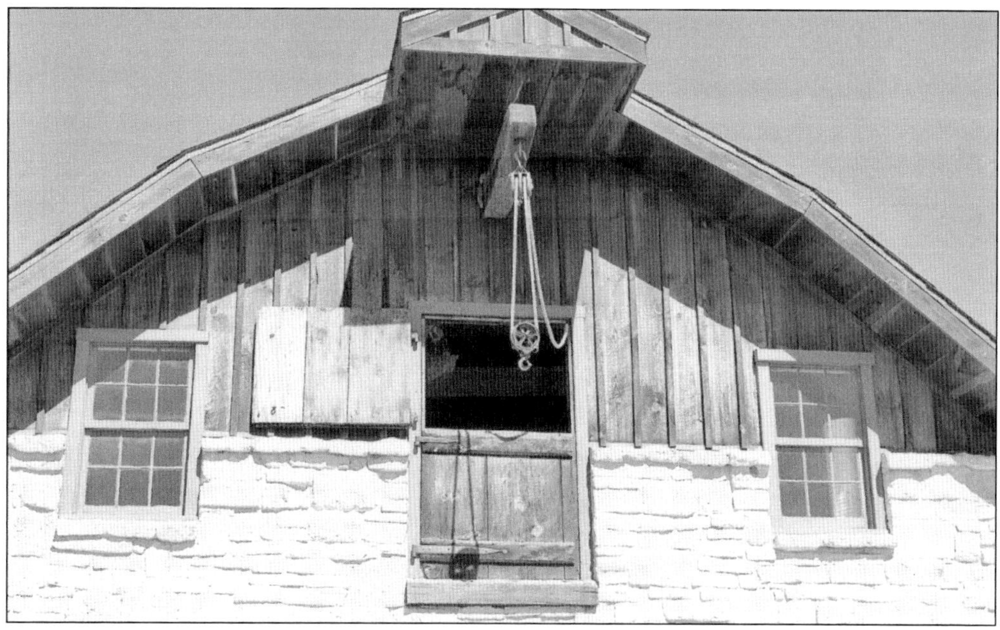

Mills such as the John Teeter Mill relied in part on gravity feed and were often two or three stories tall. Grain was hoisted in sacks to the top floor, where it was channeled from the grain bin to the mill cradle, which shook the grain into the eye of the stone. The Dutch doors allowed the top door to be opened for ventilation while the bottom remained closed. (Author's collection.)

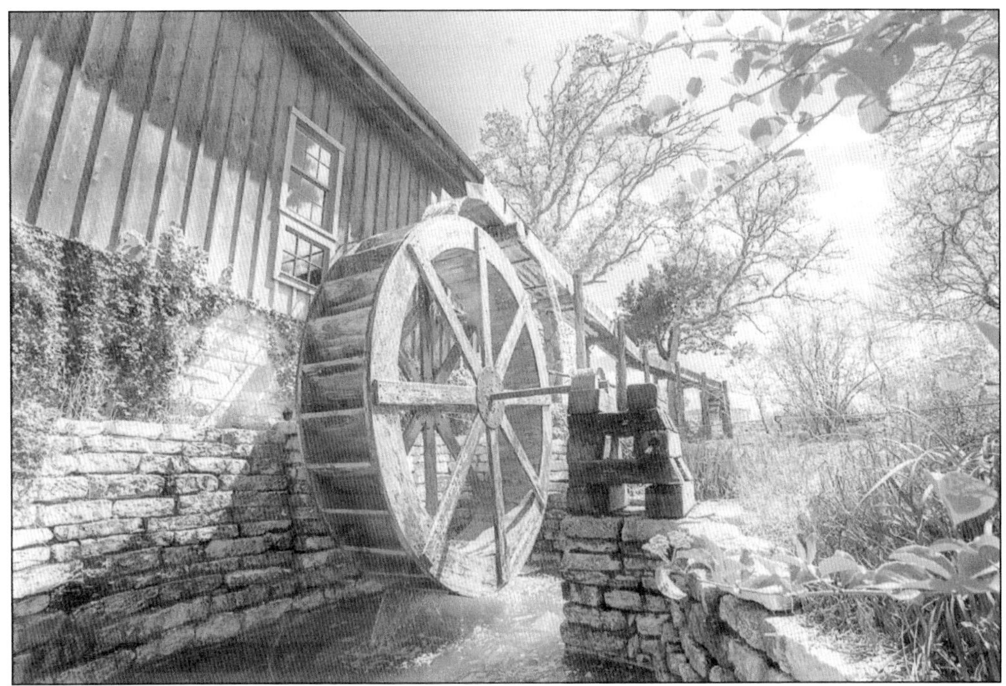

In the 1700s, when the mill was built, power was provided to the mill machinery by a shaft from the waterwheel into the basement. The mill was later converted to a turbine operating from an underground stream, which allowed for milling through the winter months. The waterwheel and flume shown in this photograph were handmade by the Homestead Heritage Village craftsmen to restore the mill to its original power arrangement. (Ben Owen.)

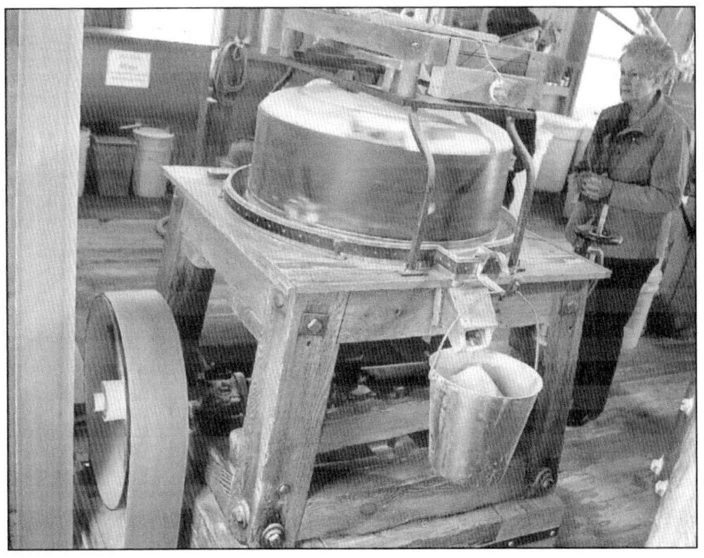

The newly restored mill uses a set of French burr stones that weigh 500 to 700 pounds each. Each millstone has a flat area called land and a special design of grooves called furrows. As the top stone turns, it tears or cuts the grain rather than crushing it. The grooves on the bottom stone direct the flour or meal to the outside of the wheel, where it falls into a container attached to the stand. (Author's collection.)

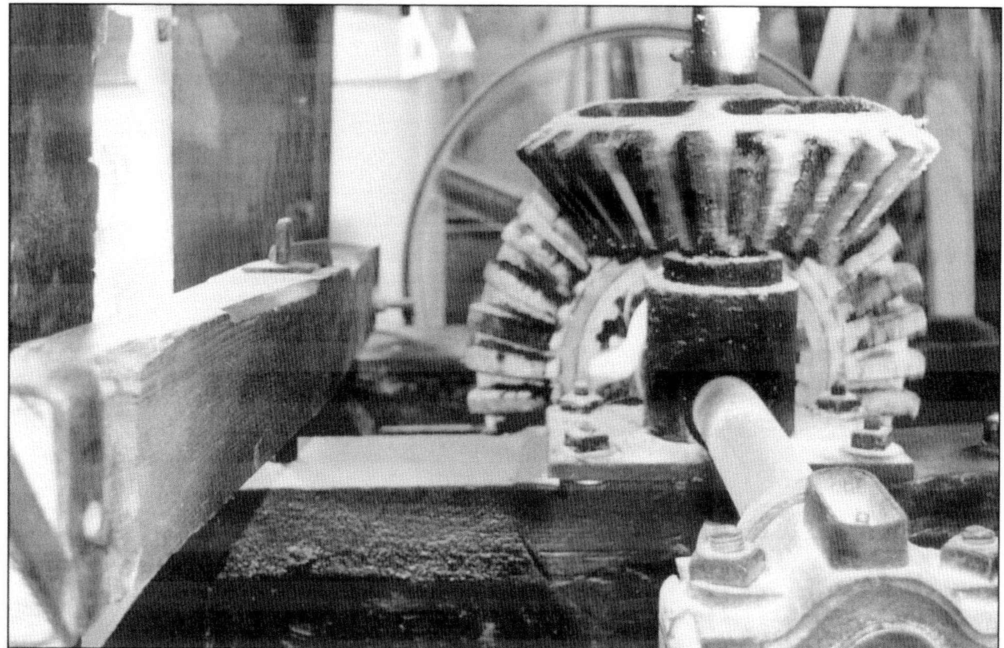

This view from underneath the millstones shows a large wheel with a peripheral belt that brings power for the millstone up from the basement. A bevel gear mounted to the horizontal shaft is mated with another mounted to the vertical shaft. This shaft rises through a hole bored through the axis of the stationary lower bedstone to turn the upper runner stone. (Author's collection.)

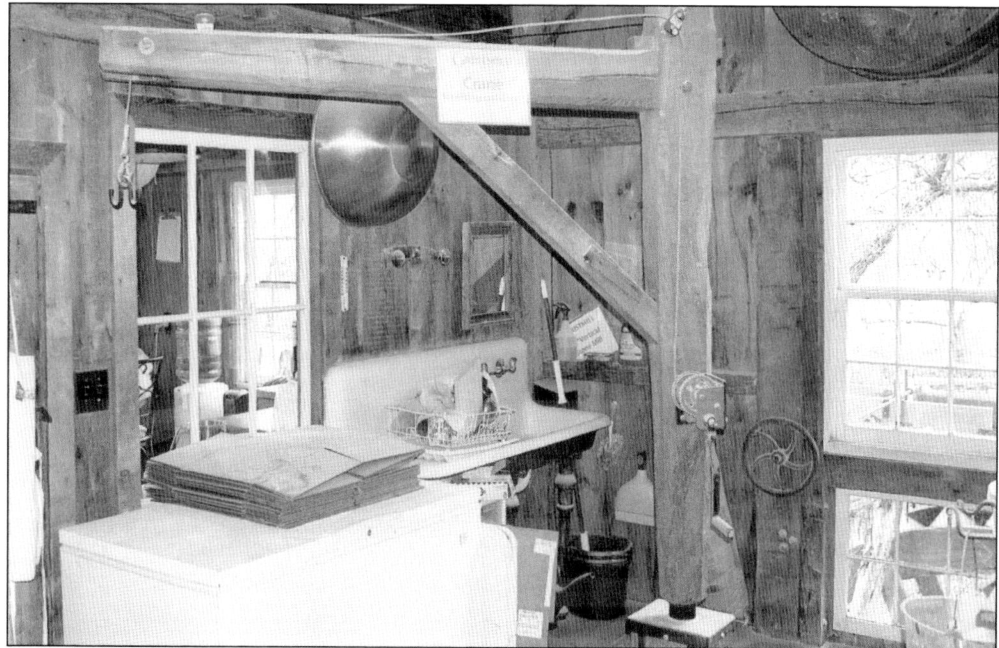

This crane is made of oak and is used to lift the runner stone off the bedstone so the stones can be flipped over and dressed or sharpened. As the stones grind, they wear down and become dull. The distance between the two stones is set according to the type of grain being ground. Wheat requires less space for grinding than corn. (Author's collection.)

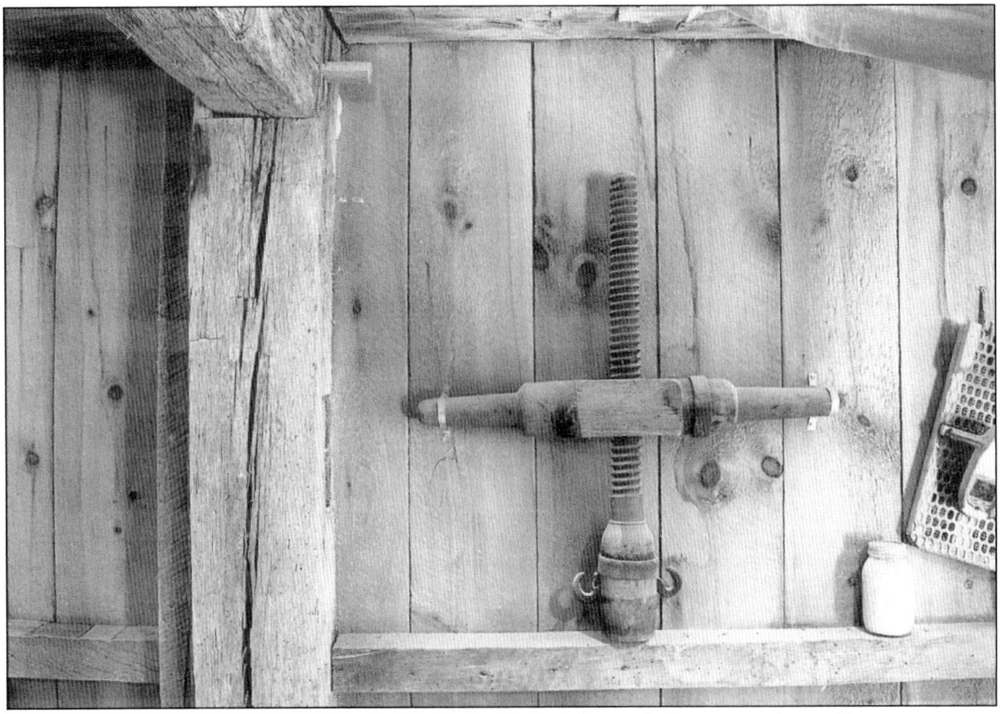

The tool shown on the wall was attached to the crane and assisted in lifting the stones for maintenance. This device with wooden threads was original to the mill. It usually took two or three days to dress a pair of millstones. The dressing was done as needed, depending on the volume of grain that passed through the stones. Millstone dressing was normally done by skilled millers. (Author's collection.)

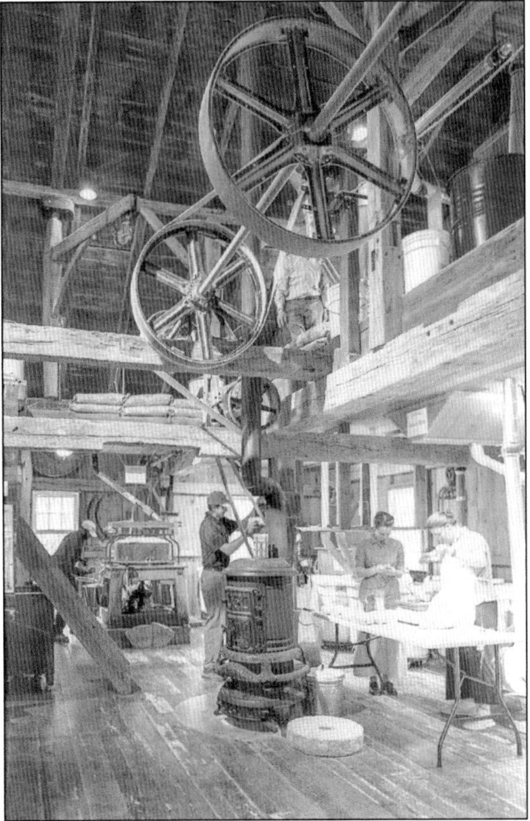

The upper belts and pulleys were acquired from an old Texas cotton gin and repurposed to operate three of the four mills in the building. Those working the mill are Danny Seifu, at the stone; Brian Claborn, near the stove; and Shahar Yarden, walking in the loft. Working at the table are Batyah Yarden (left) and Cathy Miller. The Homestead Gristmill is one of two working mills in Texas. (Ben Owen.)

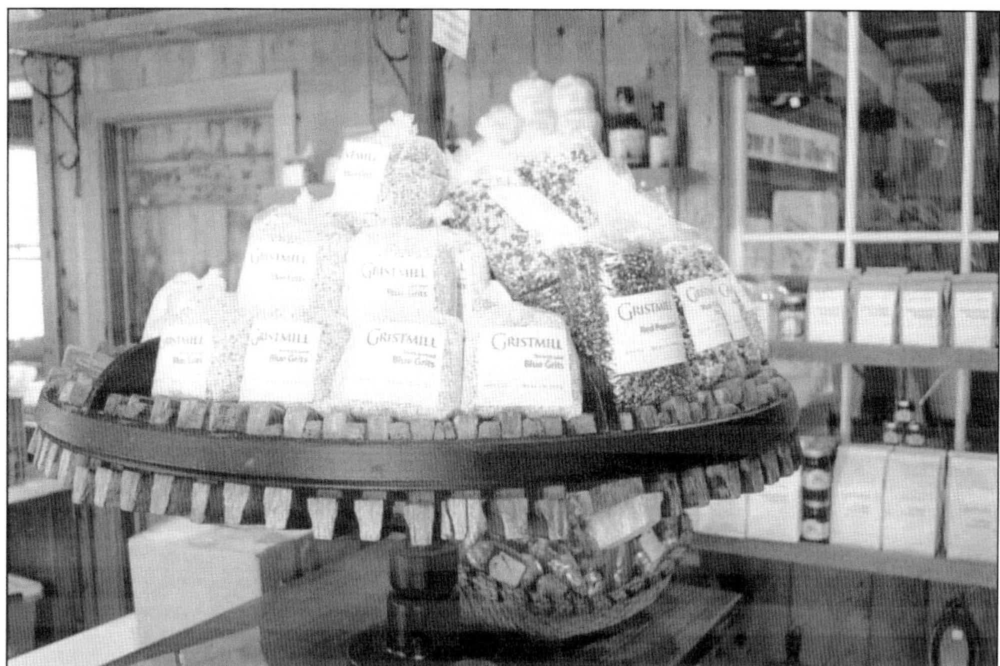

One of the original gears from the Teeter mill is used for displaying flour and other products made at the mill. Notice that this gear has wooden teeth. The wooden teeth were more practical than the iron-toothed gears because if an iron tooth broke, the entire gear would have to be replaced. The teeth were individual elements separated by spacer blocks, so a worn or damaged wooden gear could be easily replaced. (Author's collection.)

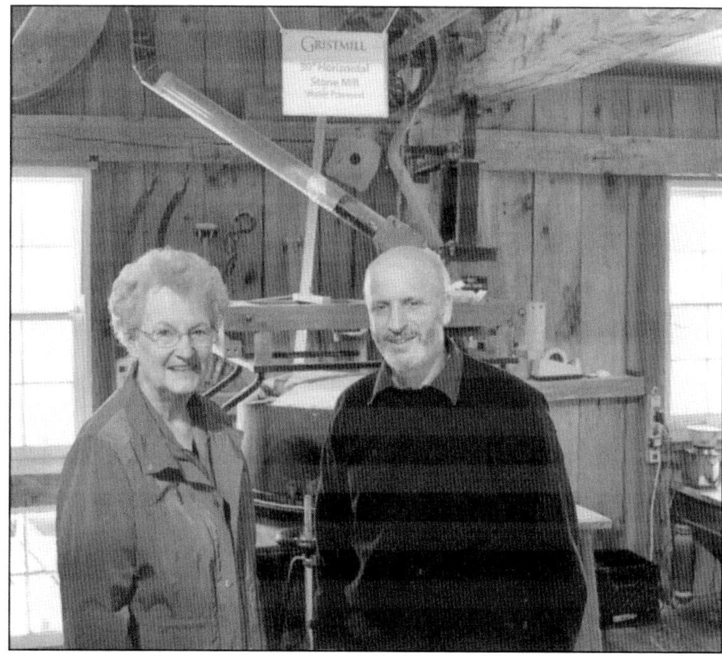

Author Charlene Carson, left, and miller Shahar Yarden stand next to the millstone. Yarden explained that a miller can often tell if the mill is working properly just by listening to the rumblings of the wheels and gears. Homestead Heritage Village, an agrarian, craft-based, Christian community, is open year-round. Visitors are invited to watch and participate as the craftsmen practice the crafts and homesteading skills of their forefathers. (Author's collection.)

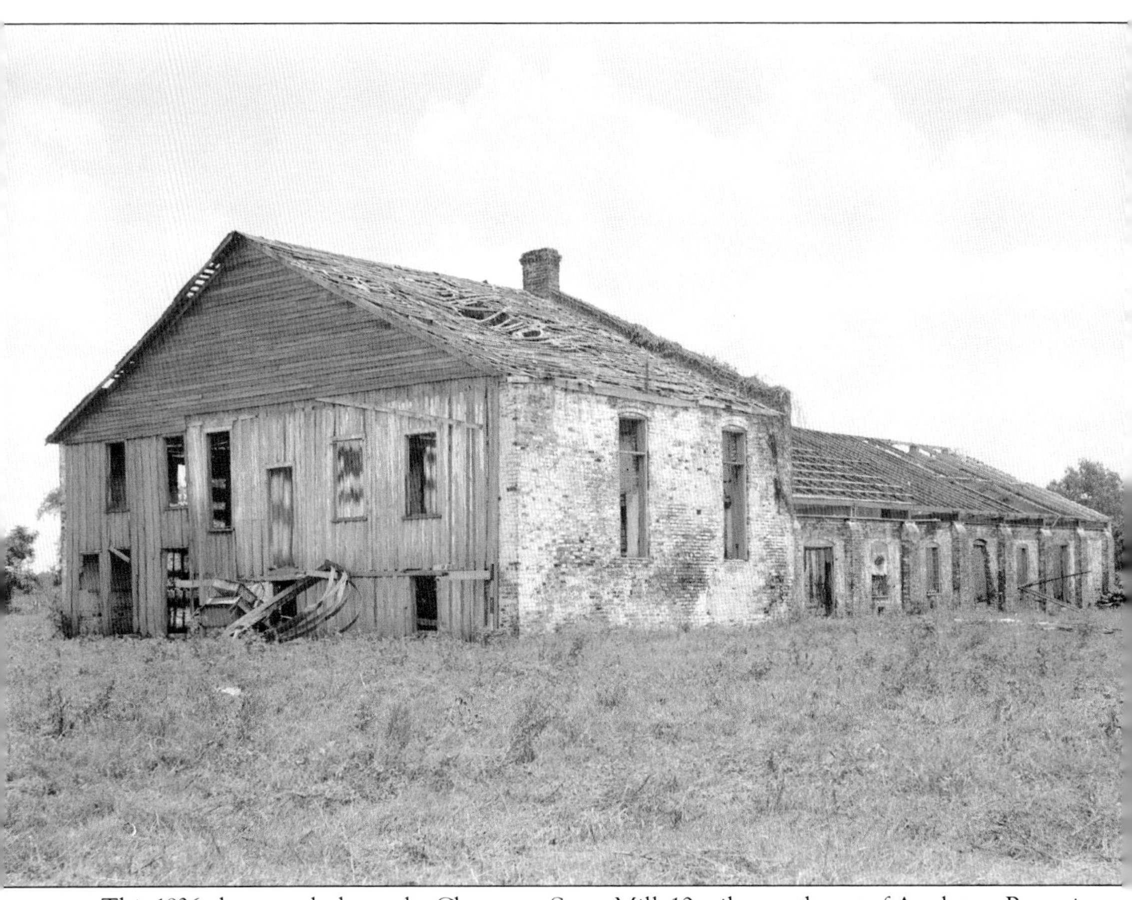

This 1936 photograph shows the Chenango Sugar Mill, 10 miles northwest of Angleton, Brazoria County. This sugar mill, built by slaves, was part of an extensive plantation established in the 1830s. Monroe Edwards purchased the mill with money supplied by prominent New Orleans merchant Christopher Dart, and the two became business partners. Edwards then used the property for a slave-smuggling operation. Edwards made frequent trips to Cuba and Africa, where he bought hundreds of slaves and brought them back to Chenango, the base of his operations. Edwards then sold the slaves to Texan and Southern planters. This venture was a huge success, but eventually Edwards and Dart had a falling out over a failed business transaction in which the court ruled in favor of Dart. With the court's ruling, Dart came into possession of Chenango, but due to lawyer fees and other expenses, the property soon passed from his hands. Over time, the plantation buildings fell into ruins, and they were ultimately leveled. (Library of Congress, Prints & Photographs Division, HABS TEX, 20 CHEN, 1-1.)

Five

The Salado River
A Hometown Treasure

The Salado River, now called Salado Creek, rises in two forks in northwestern Williamson County, unites in southern Bell County, and flows 35 miles northeastward to join the Lampasas River in the area known as Three Forks. At that point, the Salado, the Lampasas, and the Leon Rivers all unite to form Little River. From 1848 to the 1957, the Salado River of Bell County was the power source to the eight mills along its banks.

The name *Salado*, given to both the town and the creek, means "salty" in Spanish, even though the water is fresh. The river, now a treasured creek that flows through the heart of the village year-round, was the site of camping grounds for the first Indian tribes in the region. In 1838, a military road extended across the Republic of Texas and crossed Salado Creek at the place where Main Street does today. This road later became an overland stage route and was used as a branch of the Chisholm Trail.

During the Civil War, the Salado was the power source for two mills. The first was Chalk Mill, built in 1848. If the Chalk Mill was like many other Texas mills operating during the Civil War, it provided flour or other provisions for the Confederate troops. The Davis Mill, built in 1864, prepared wool for Confederate uniforms.

After the Civil War, Texas was initially under military rule. However, this did not prevent an influx of people from coming into the state. The millwrights who came into Bell County were proactive in establishing a milling business. The steady immigration of new settlers into the area fostered a need for lumber, cornmeal, flour, and other milling services. The mills along Salado Creek were there to meet those needs. Those mills, in addition to the Chalk and Davis Mills, were the Col. Thomas H. Jones Mill, Stinnett Mill, Ike Jones Mill, Dulaney Mill, Summers Mill, and Mote Smith Mill. Salado Creek has always played a significant role in the development of Salado, and in 1967 it was designated the first recorded Natural Landmark in Texas.

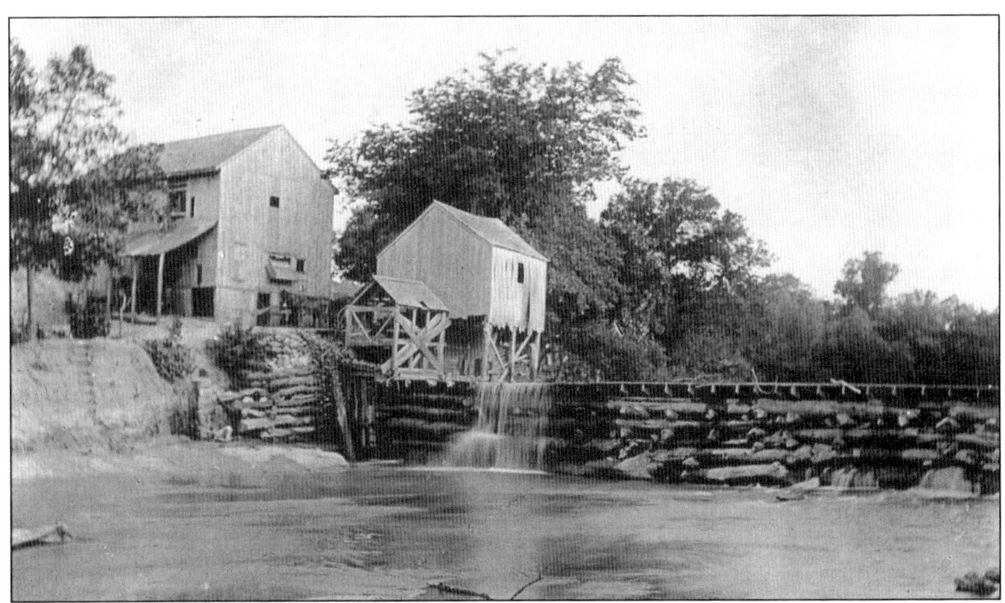

Chalk Mill, built in 1848 by brothers Whitfield and Ira Chalk, was the first of eight mills built on the east side of the Salado Creek dam. It began as a sawmill and was later equipped as a gristmill. Rev. James Ferguson acquired the mill in 1867 and operated it until his death in 1876. Ferguson's family operated the mill for several years, but it was abandoned after the flood of 1900. (Lena Armstrong Library, Belton.)

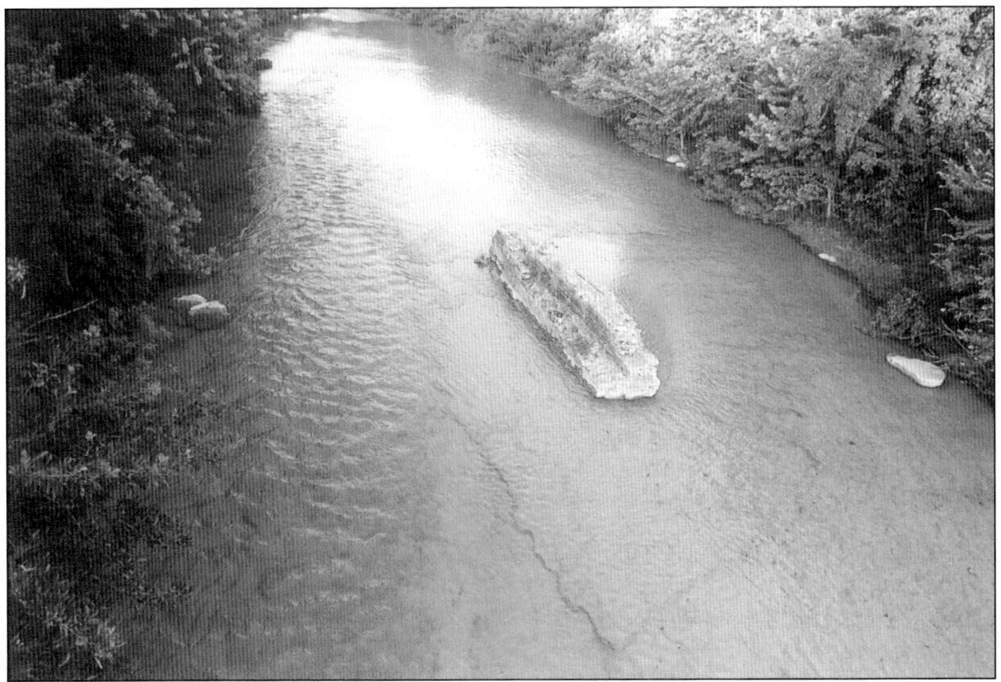

Almost 170 years after it was built, the remains of the Chalk Mill dam are still visible in Salado Creek. The dam was built cribwork style, with cedar timbers placed horizontally like a log crib. The crib was then filled with stones, gravel, and concrete that gave a semblance of permanence. (David Carson.)

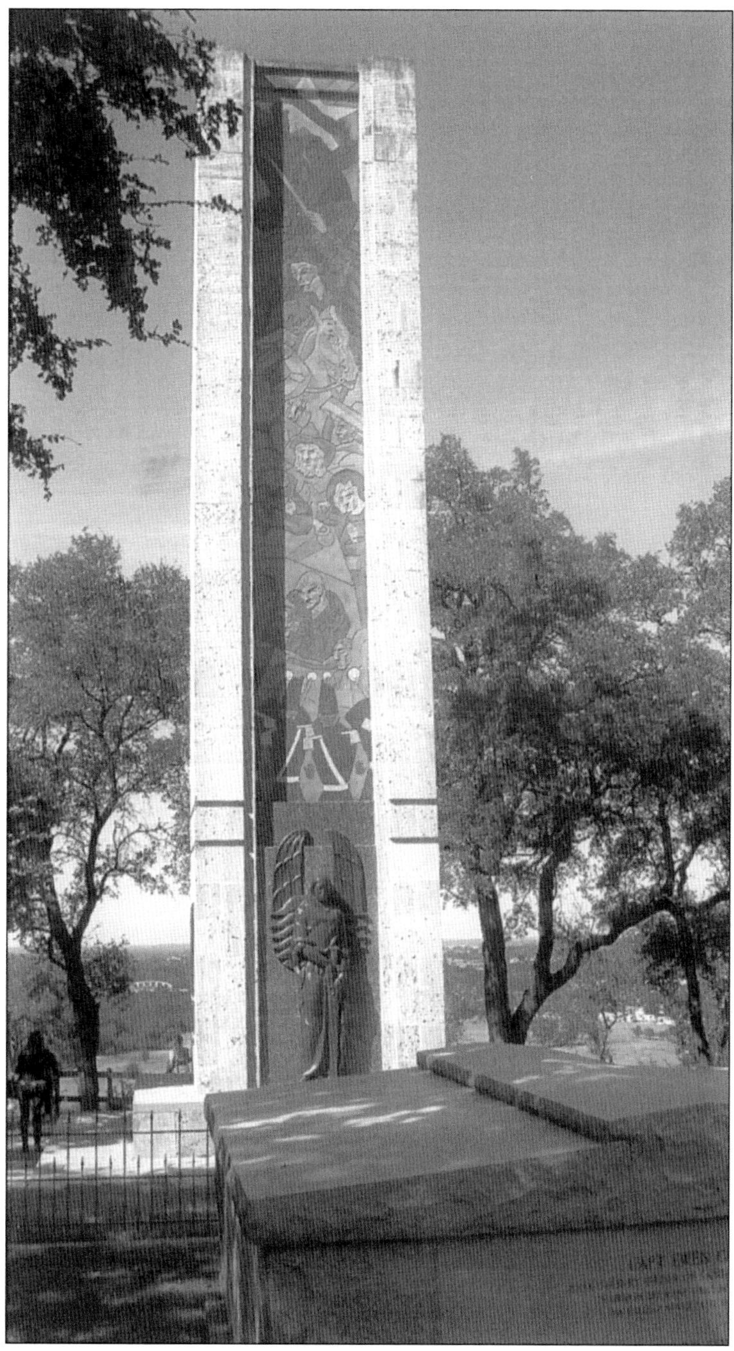

Whitfield Chalk immigrated to Texas from Tennessee in 1833 at the age of 28. He served with the armies of the Republic of Texas, fighting in several major campaigns including the ill-fated Mier Expedition of Christmas Day 1842. The remains of the men who perished in the struggle for Texas independence are entombed in a granite crypt—their names etched in stone and marked by this towering monument at Monument Hill at LaGrange, Texas. The monument is a memorial to the men who died in the Dawson Massacre and the infamous Black Bean Death Lottery following the Battle of Mier. (Author's collection.)

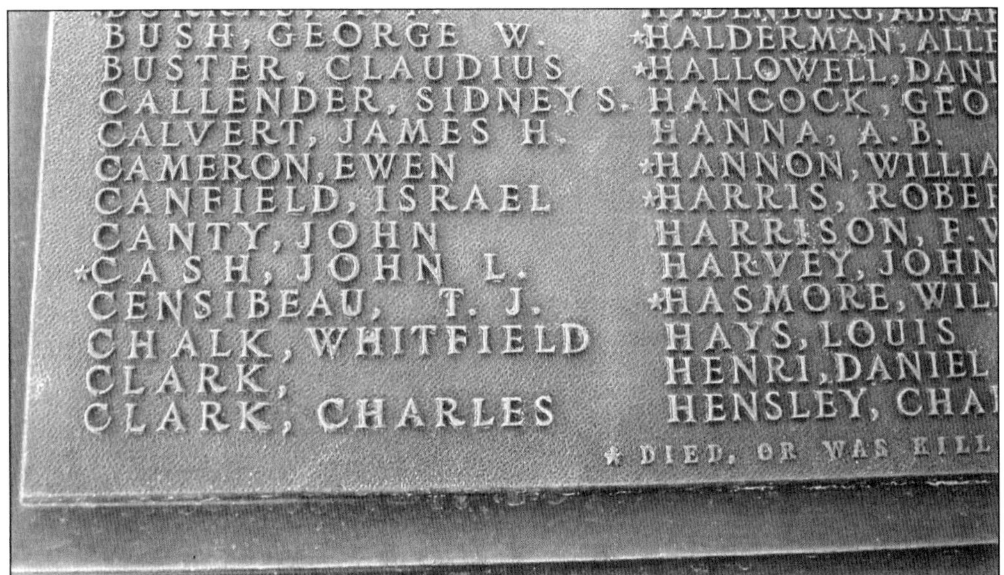

The name CHALK, WHITFIELD is inscribed with other soldiers who fought in the Battle of Mier. Even though Chalk escaped, his name is inscribed on a bronze plaque on one side of the monument tower. Also, for services rendered to the struggling Republic, President Sam Houston, on August 5, 1844, commissioned Chalk as a Major of the Second Regiment of the First Brigade of the Militia of the Republic of Texas. (Author's collection.)

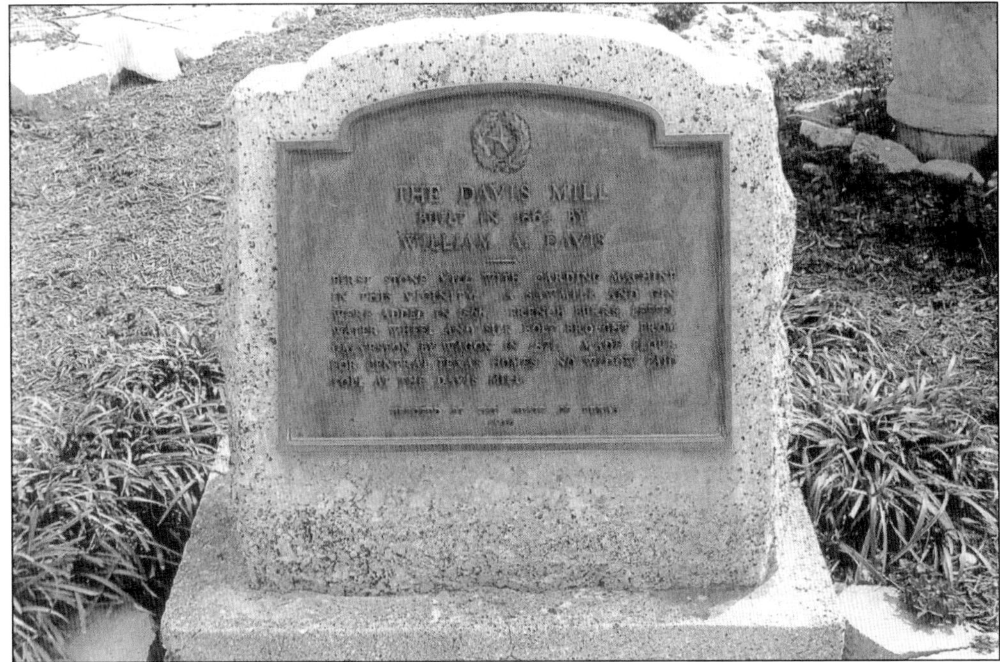

This marker commemorates the William A. Davis Mill, which was built at the headwaters of Salado Creek in 1864. Davis installed a wool carding machine to prepare wool for the manufacture of Confederate uniforms. After the war, Davis converted his mill to a gristmill. He added burr millstones imported from France, a Leffel waterwheel, and silk bolting to clean the grain and sift the flour or meal to the desired fineness. (Author's collection.)

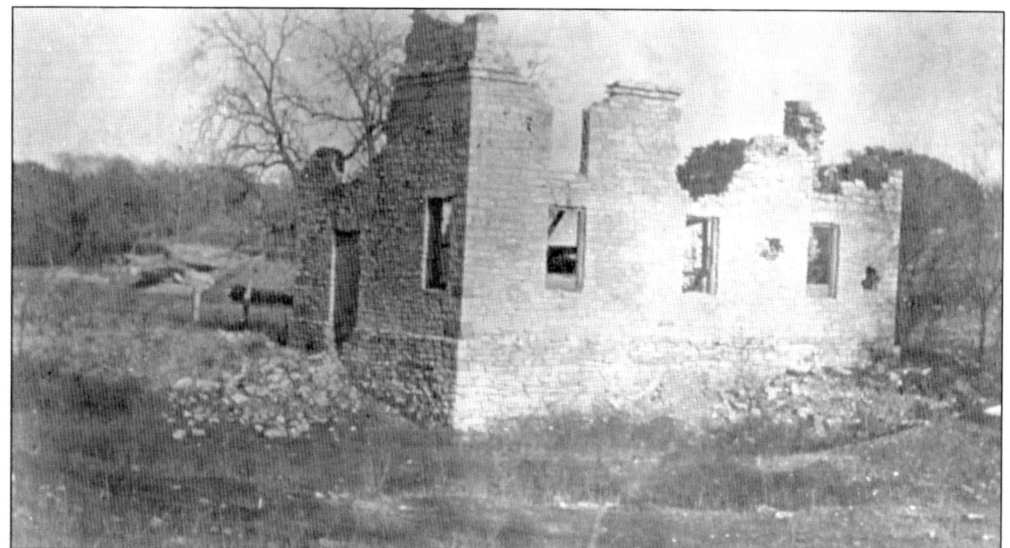

Davis operated the mill for 24 years before selling it to William A. Pace, formerly of Georgetown. The mill was in operation until 1900, when heavy rains during the Galveston storm caused Salado Creek to go on a rampage, demolishing everything in its path. The mill sat abandoned for years until the flood waters of December 1913 washed away its final remnants. (Sophia Vickrey Ard.)

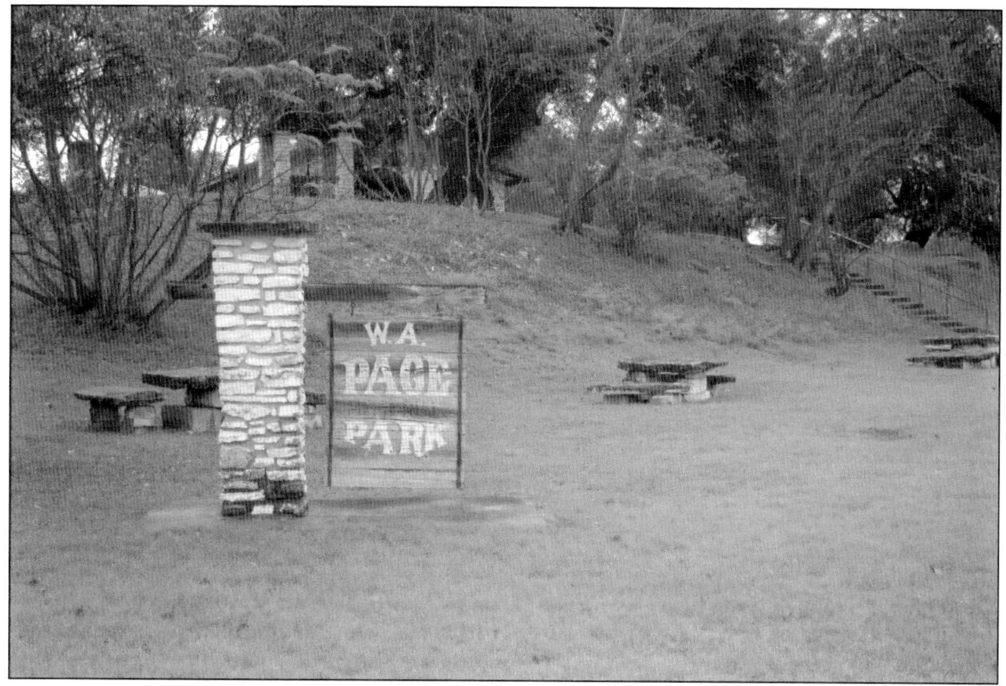

In August 1946, Pace heirs John Pace and his sister Elizabeth (Mrs. John S. Hodges) offered the land to Salado for a park, provided it be named W.A. Pace Memorial Park. Because of their generous donation, Salado now has a beautiful park on the north bank of Salado Creek, the former site of the Davis Mill. (Author's collection.)

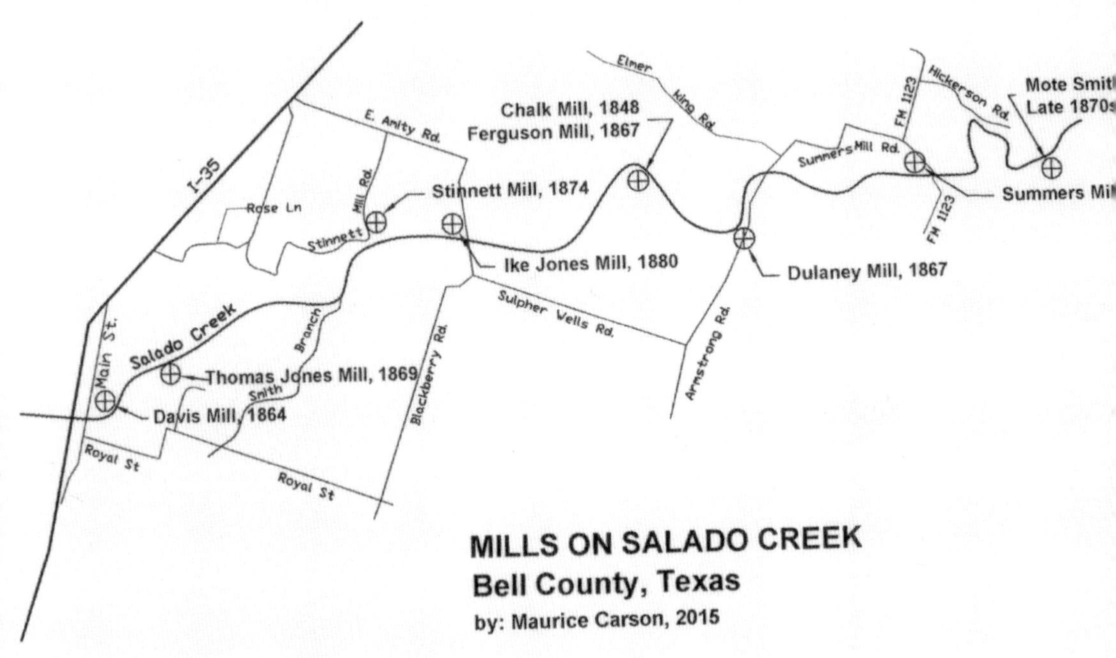

Because of its strong waterpower, the Salado River, as it was referred to in early writings, had more mills than any other stream in Bell County and made the county an industrial center. This drawing shows the location of each mill and the year it was built. The first mill in operation was Chalk Mill, built in 1848. The last mill in operation was Summers Mill, which ceased in 1957. Although only nine miles long and not so deep as other streams, the Salado attracted millers because of its never-failing water supply. The distance between the mills ranged from one mile to two and a half miles. The dams were built just far enough apart to allow the water released from one mill to gain sufficient volume before it had to turn the wheel of the mill below it. (Maurice Carson.)

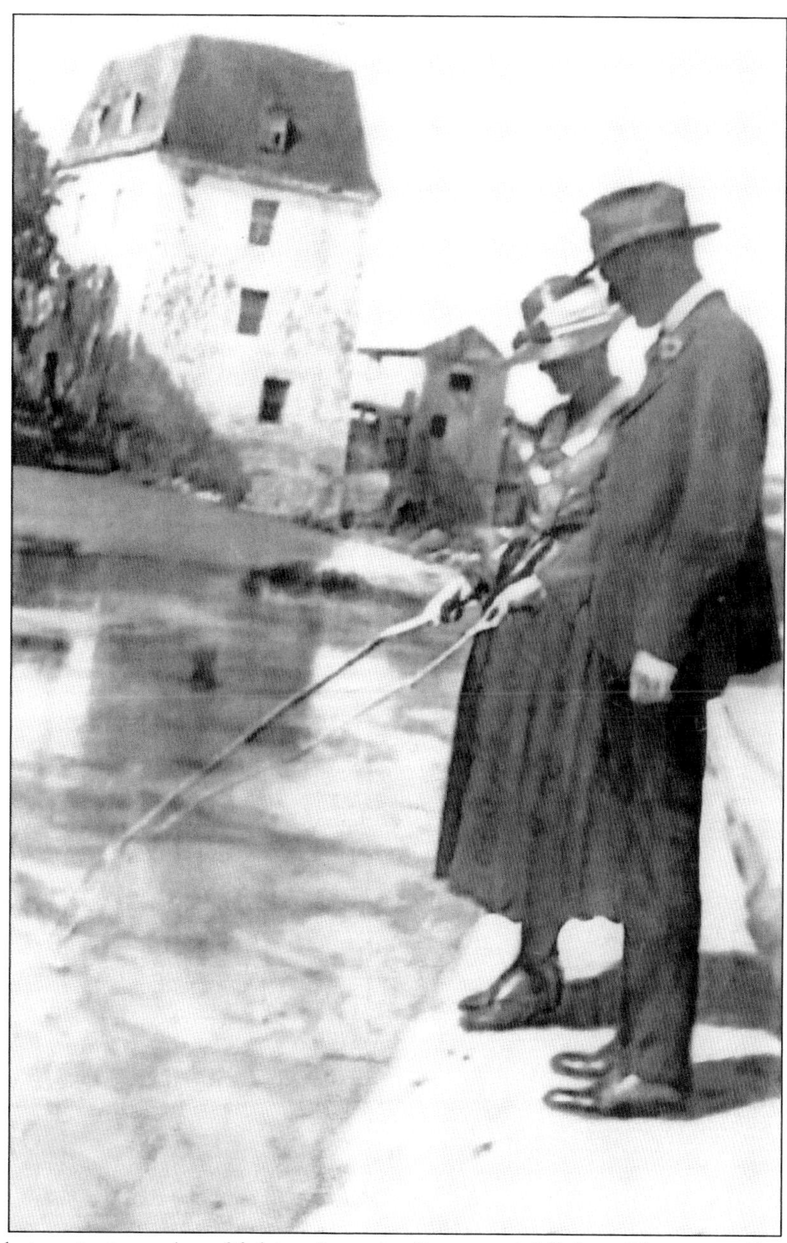

This couple is enjoying a day of fishing from the Summers Mill dam. The gristmill, seen in the background, was built soon after the Civil War by Bell County pioneer Col. John Meyers. The waterwheel was a Leffel undershot turbine, which was installed inside the millhouse. Summers Mill has never had any type of waterwheel on the outside of the millhouse. The millstones, which had come from France, the turbine waterwheel, and other machinery was hauled up from the Texas coast by oxcart. Myers operated the mill for almost 15 years before D.C. Summers acquired it. When Summers took ownership in 1879, the mill and community that grew up around it became known as Summers Mill. Summers installed modern machinery, making it the first roller mill in the county. Millstones for grinding had been in use almost since the beginning of milling. Replacing them was a giant step forward. Soon millstones would be a relic of the past. This photograph shows Summers Mill as it appeared prior to the 1930s. (Kelsey collection.)

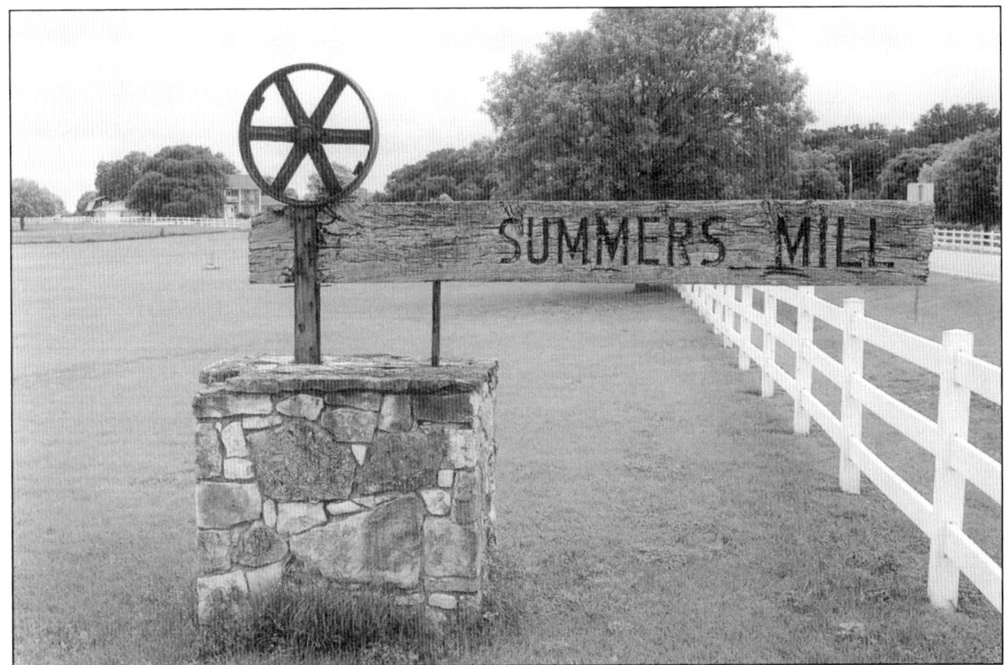

By 1884, the Summers Mill community had a population of 50. There were three churches, a school, and the mill, which in addition to grinding corn had added a flour mill and a sawmill. The sign post that greets visitors today is topped with one of the drive-belt pulleys from the era when Summers Mill was an industrious farming community. (Author's collection.)

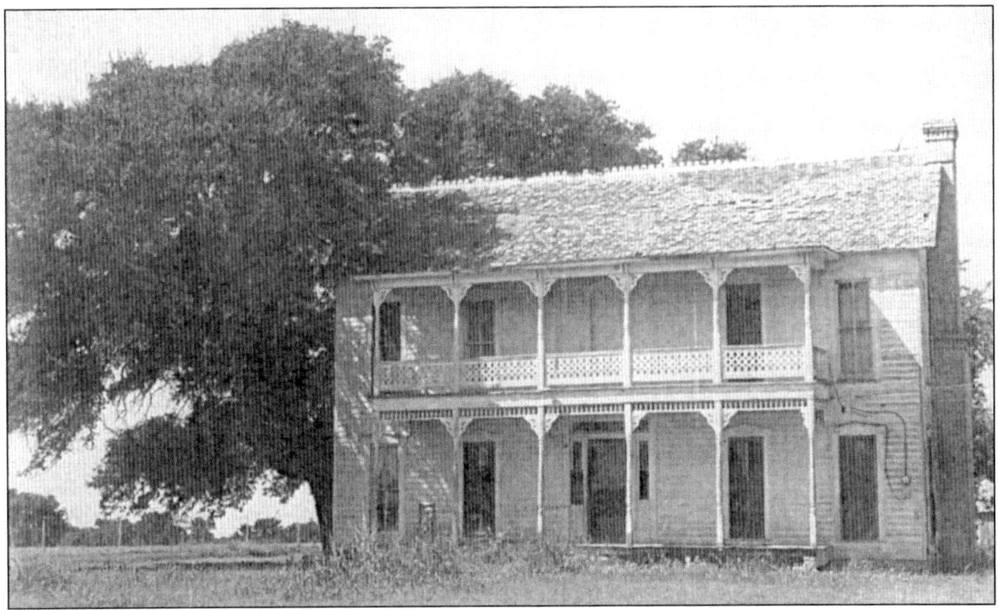

D.C. Summers operated Summers Mill for 10 years before he traded it to J.R. Holland for 226 acres of farm land in 1889. The Holland family had come to Bell County from Arkansas in 1874. Holland built this home at Summers Mill in 1894. This stately home, with its ridge row roof and porches with latticework bannisters and cornices, became the showplace of the community. (Lena Armstrong Library, Belton.)

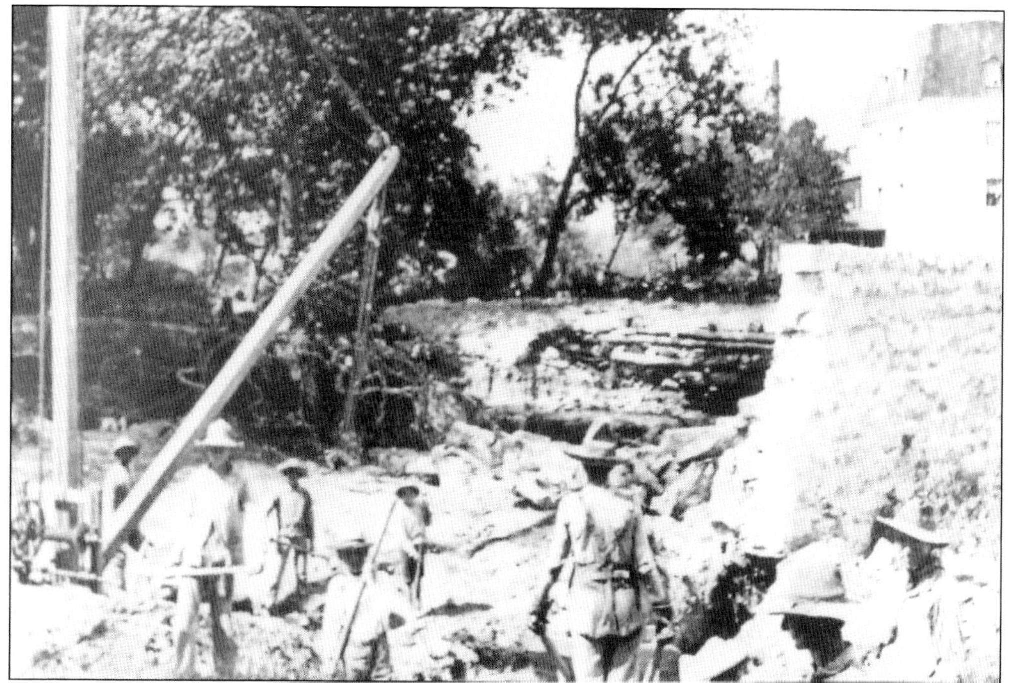

When J.R. Holland took ownership of Summers Mill, he made several changes to the milling operation. He replaced the original cedar post dam with a more permanent structure made of stones and concrete. Above, men work on the dam. With cotton becoming a more productive crop in Bell County, Holland also installed a gin at the mill. The photograph below shows workers at the cotton gin. Both photographs are from the 1880s. When Holland acquired the Summers Mill property, he purchased the nearby 70-acre pecan grove. The grove became the meeting place for political rallies, camp meetings, family reunions, and moonlight picnics. The nearby Salado Creek served as a place for baptisms and swimming. It was understood that there would be no swimming on Sunday, especially if there was to be a baptism. (Both, Summers Mill Archives.)

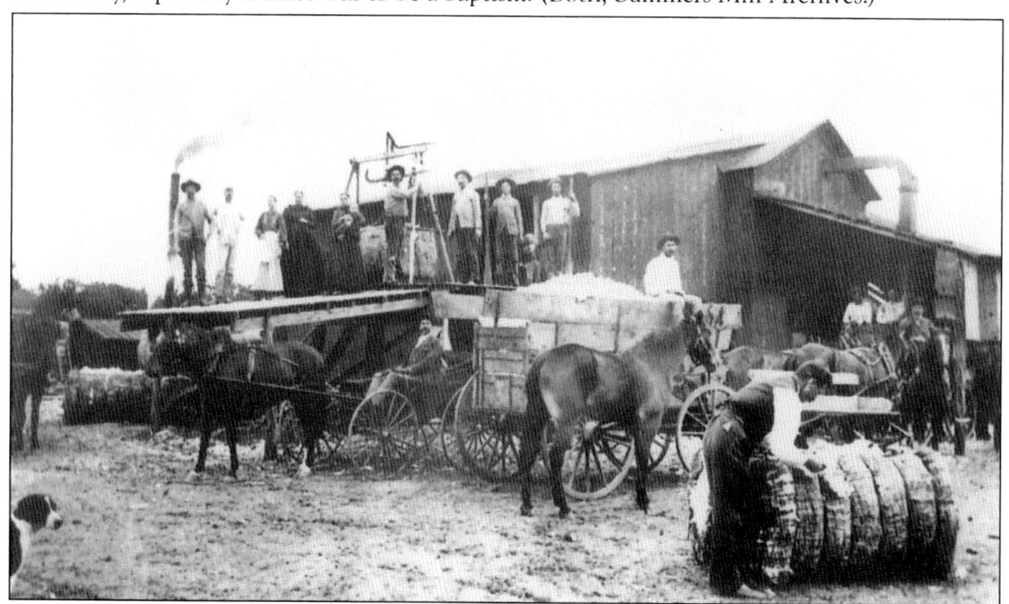

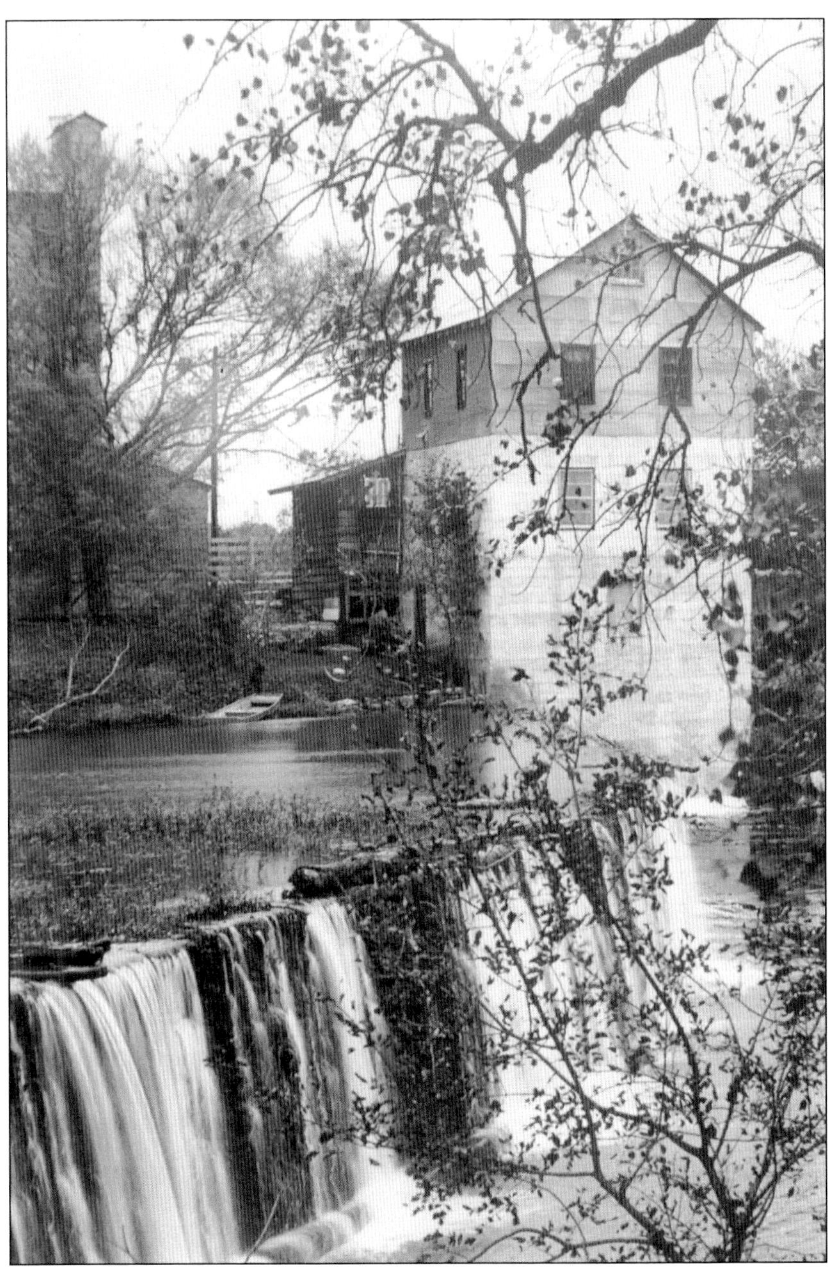

After Holland's death in 1912, his son-in-law J.M. Phillips continued to operate the mill. During his ownership, the mill was completely destroyed by the flood of 1921. It has been reported that the Salado River rose over three feet within 20 minutes, and still the rains came. Even after this devastating setback, in 1932, Phillips and his son Leland rebuilt the mill. Phillips made a few changes in the appearance of the mill building—the most noticeable is the change in the roofline. The original mill had more of a mansard-style roof, while the new mill has a gable-style roof. The new mill also has more windows than the original building. People were skeptical about the new mill's success, but 20 years later, it was still turning out 15,000 pounds of fresh cornmeal in a 12-hour day. In 1966, Phillips sold the mill to Jimmie and Pearl Haddon, and they converted it into a restaurant, which they operated into the 1970s. (Summers Mill Archives.)

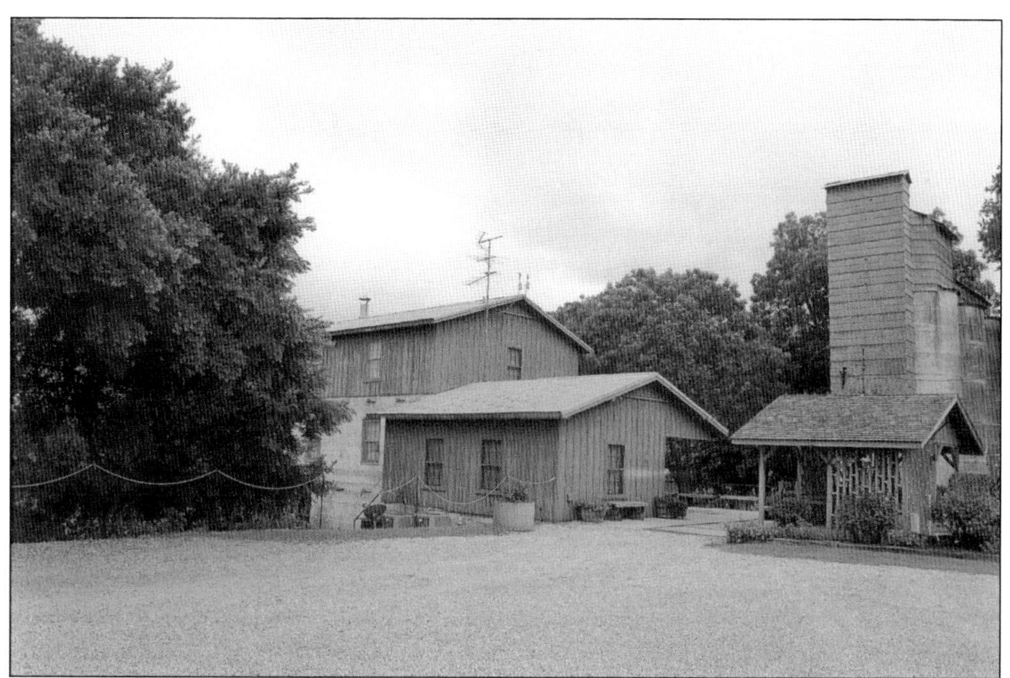

These two photographs represent Summers Mill as it is today. The mill building is the same one that J.R. Phillips built in 1932. Inside, it has been modified to make it into a comfortable, well-appointed home. The small building to the right has been added. However, the original scales are visible in front of it. The silos featured below are original to the site. Their size indicates the amount of grain that Summers Mill could process. The grain was augured in at the top and was then stored until it was released through the small doors near the bottom of each silo. From there, the grain was transferred to the mill by a ground-imbedded augur. (Both, author's collection.)

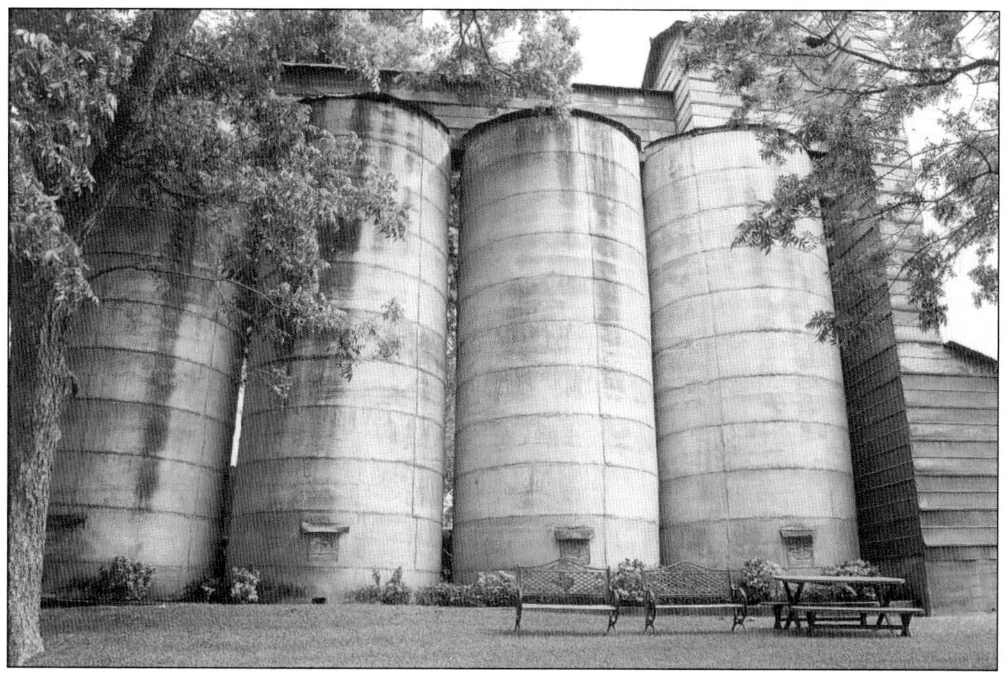

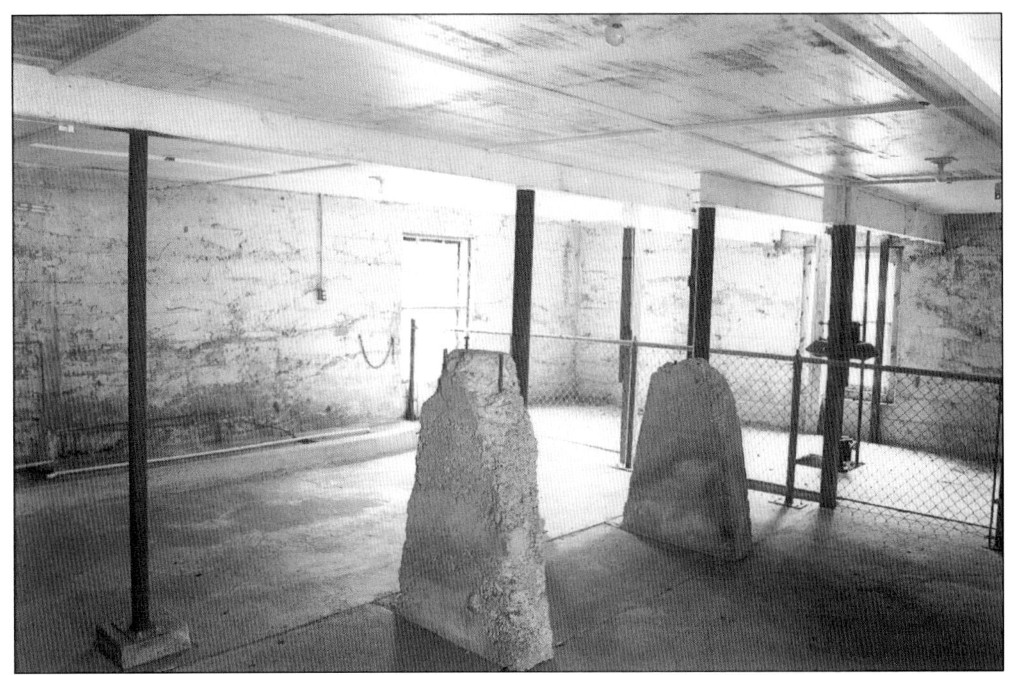

This is a view of bottom floor of the mill as it appears today. The gear sprocket on the far right is the only piece of machinery left from the 95 working years of the mill. Of the eight mills that once lined Salado Creek, two of those historic structures, Summers Mill and Stinnett Mill, still exist as a vital part of Salado and Texas history. (Author's collection.)

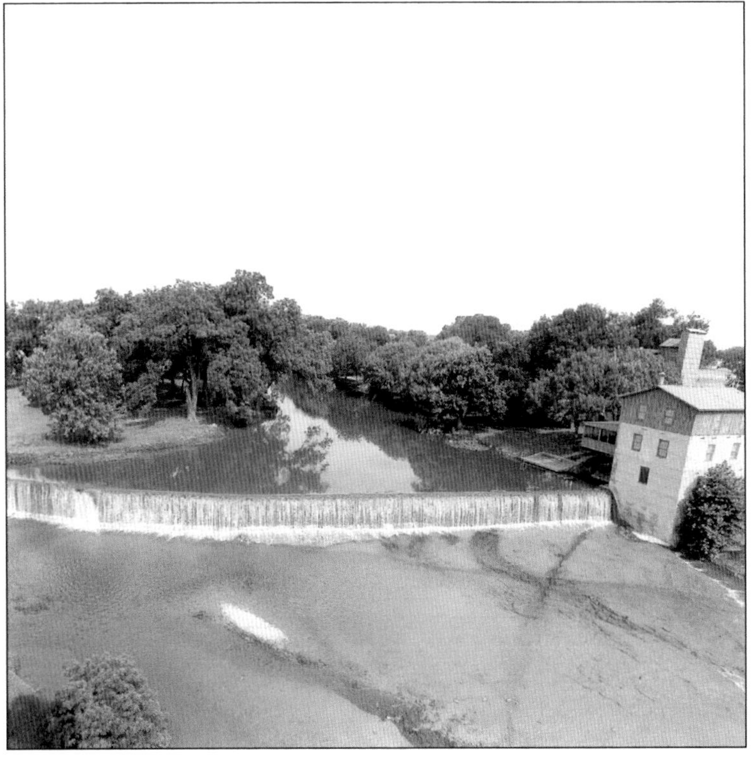

Presently, Summers Mill is owned by the Paul J. Meyer Foundation, a nonprofit charitable organization, and is known as Summers Mill Retreat & Conference Center. Facilities are available to businesses, church groups, and other organizations who are seeking a place to hold conferences and retreats in a beautiful setting designed for ministry and renewal. A Texas Historical Commission marker was dedicated at the site in 1968. (David Carson.)

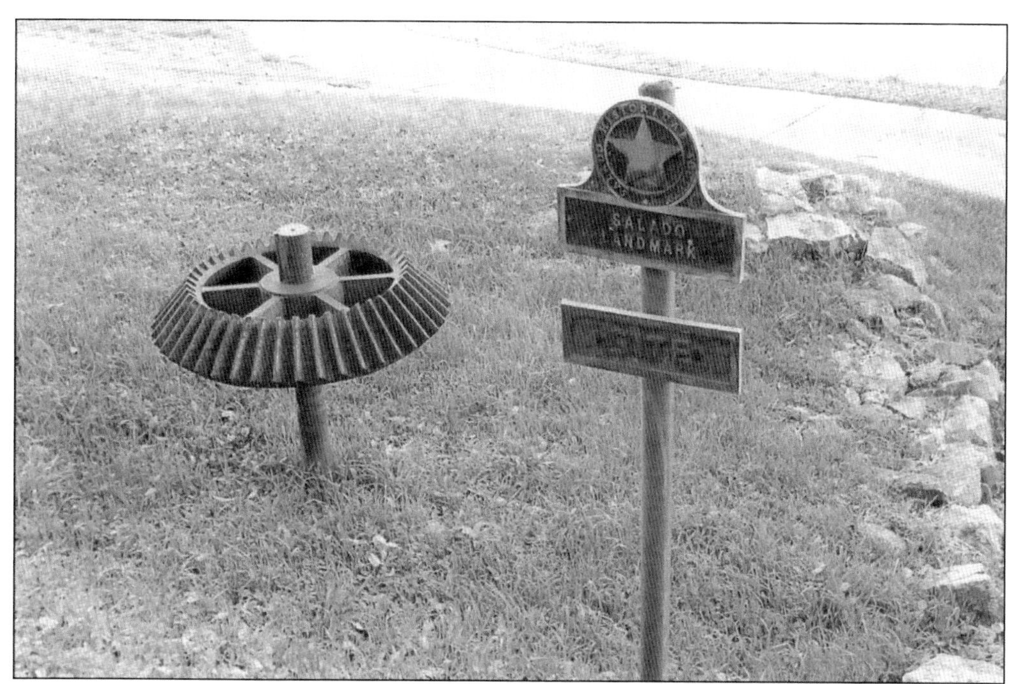

A sprocket and a portion of the original rock wall are all that remain of the Col. Thomas H. Jones Mill, which was built in 1869 and operated until 1884. Jones built it as a gristmill. After 1870, when cotton became an important agricultural crop in Central Texas, Jones was the first of the Salado millers to add a water-powered cotton gin to his mill utilizing a wooden screw-type press. (Author's collection.)

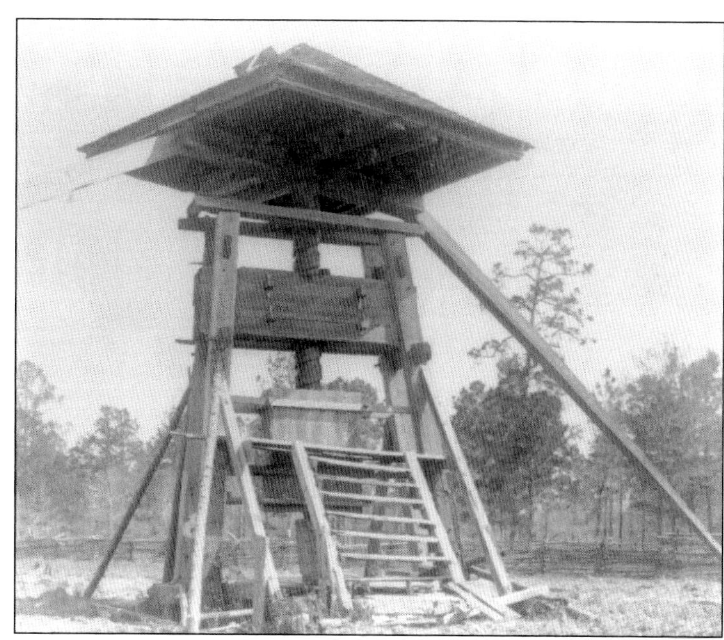

This photograph shows an example of a wooden screw-type press. After the cotton had been ginned and cleaned, workers carried it to the press, which consisted of a wooden screw with a plunger to compress cotton inside the bale box. Mules harnessed to the sweeps walked in a circle around the press to lower the screw. The pressed cotton bale weighed between 400 and 500 pounds. (Library of Congress, Prints & Photographs Division, LC-USZ62-30090.)

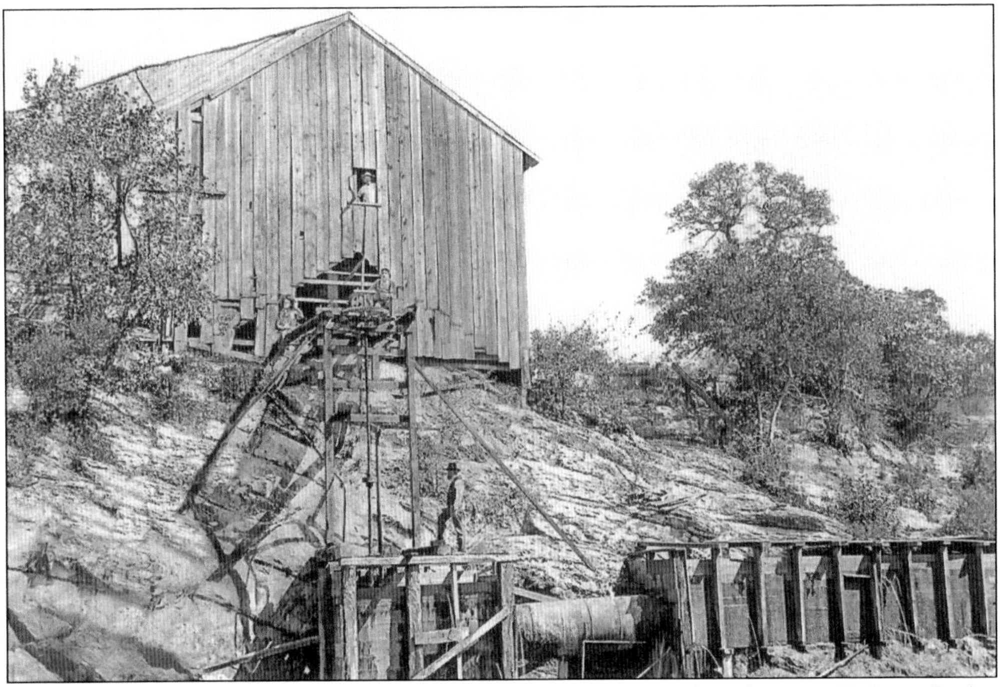

In 1867, John Thornton Dulaney built what became one of the largest and most successful of the mills on the Salado River. The mill, located near the present-day Armstrong community, operated as a gristmill, sawmill, and cotton gin. The millrace was cut out of solid rock. The cedar post dam was so well built that it withstood the ravages of floods for many years. (Robertson Plantation Archives.)

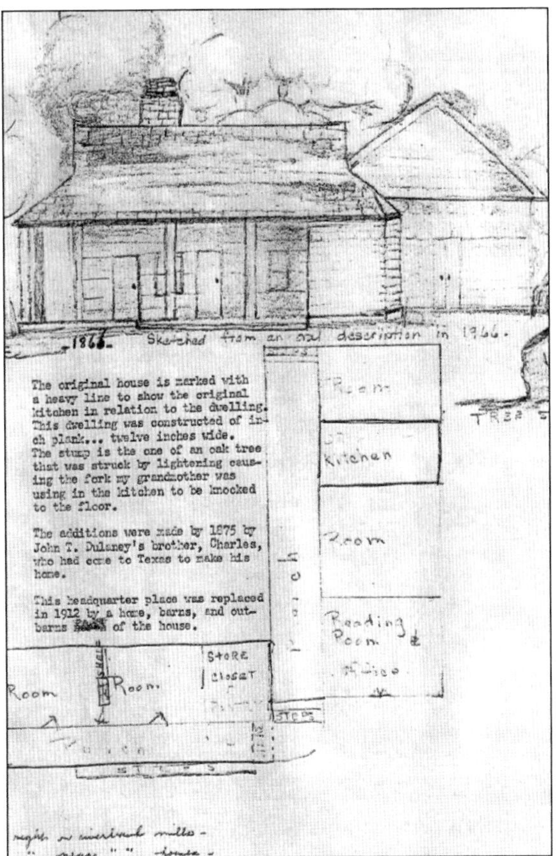

This is a sketch of the Dulaney home, located near the Dulaney Mill. In addition to the mill, Dulaney built a farm complex consisting of his home, a general store, a blacksmith shop, a church, and a school. Dulaney built tenant houses for his workers along the banks of the creek. During the peak of the working season, 25 to 30 men worked at the mill. (Central Texas Area Museum, Salado.)

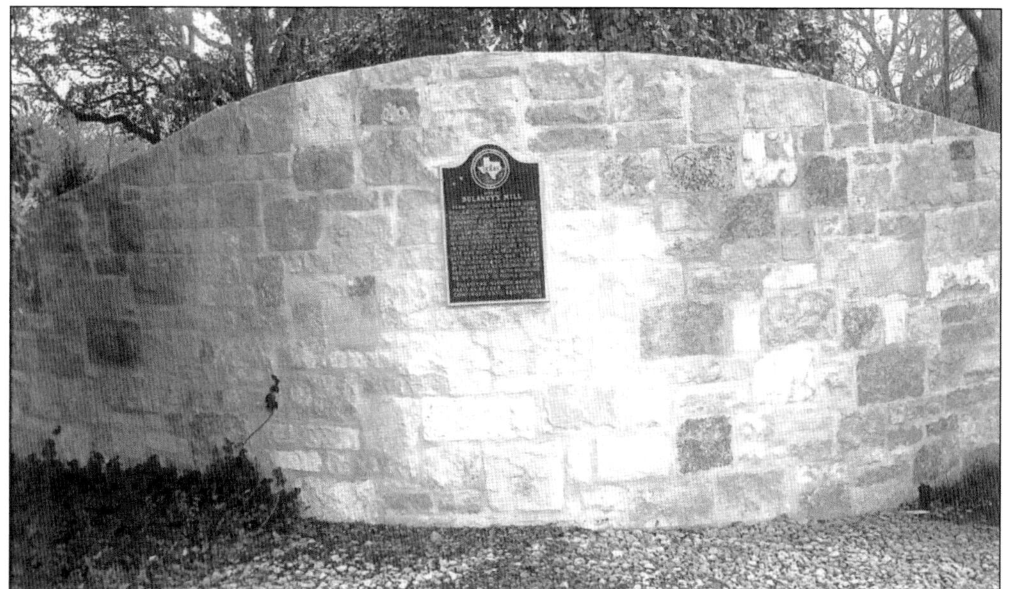

After 45 years, the Dulaney Mill ceased operation in 1912. Dulaney died in 1916 near Lexington-Fairfield, Virginia, where he was buried. His wife, Mary Jane Gates, died in 1925 and is buried in the Dulaney Cemetery on Elmer King Road near Salado. A Texas Historical Commission marker, mounted to a stonewall fence in front of a home near the original mill site, marks the significance of the Dulaney Mill. (Author's collection.)

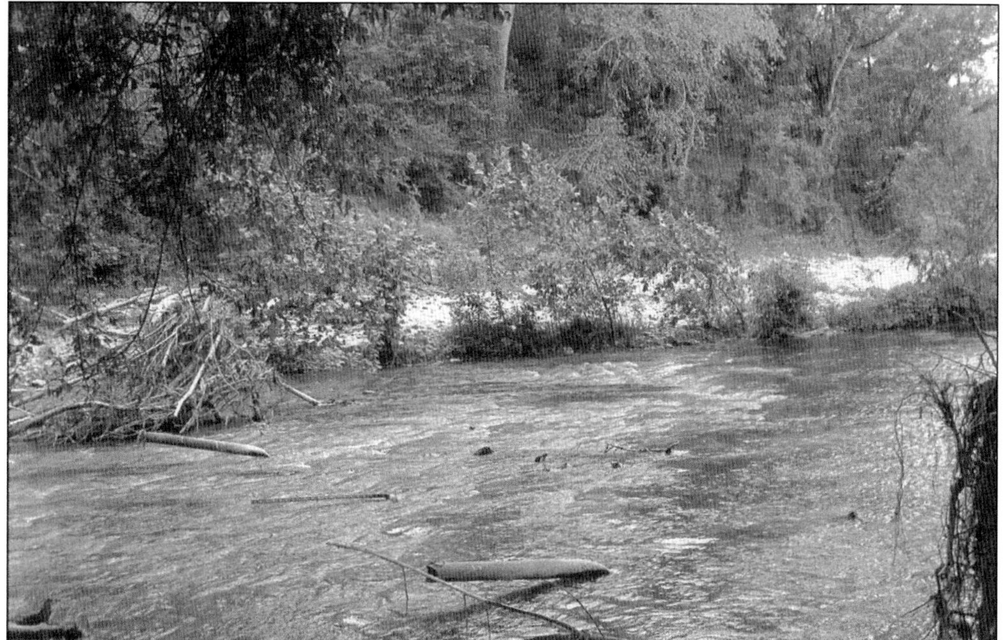

This photograph shows the site of the last of the eight mills on Salado Creek. In the late 1870s, J. Morton "Mote" Smith built a gristmill and gin along the creek bank in a pasture owned by the Garrison family of Salado. The site is east of FM 1123 near Hickerson Road. The mill was completely destroyed in 1900 as a result of heavy flooding due to the Galveston storm. (Author's collection.)

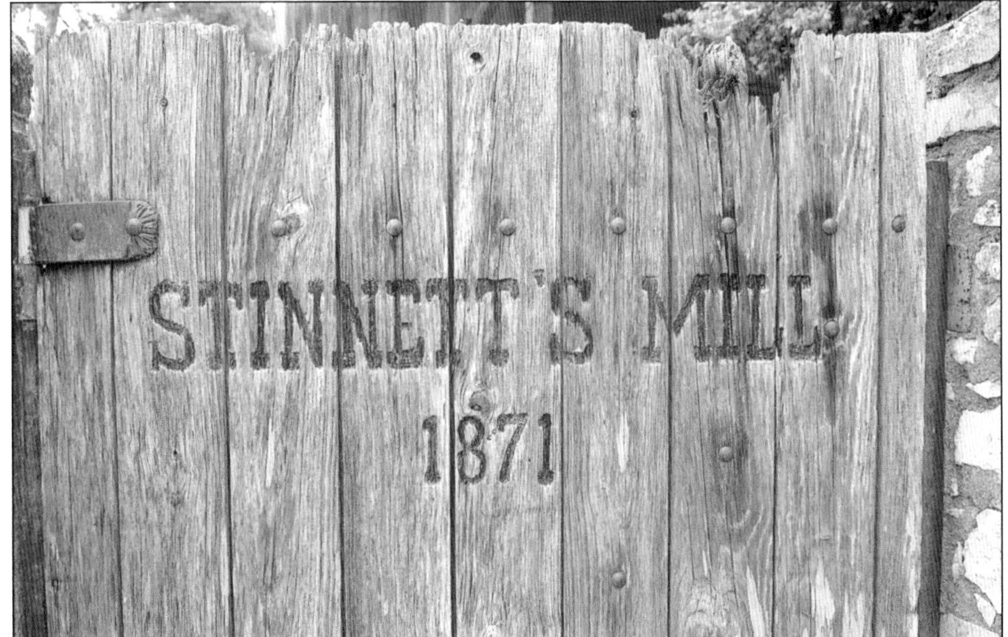

Inside this gate stands the only surviving original mill on the Salado Creek. The mill is about three miles northeast of Salado via a winding road. On the approach to the mill, the pavement stops and a dirt road begins. The dirt road is reminiscent of the days when the mill was a working mill. The road is just wide enough for two wagons meeting each other to pass. (Author's collection.)

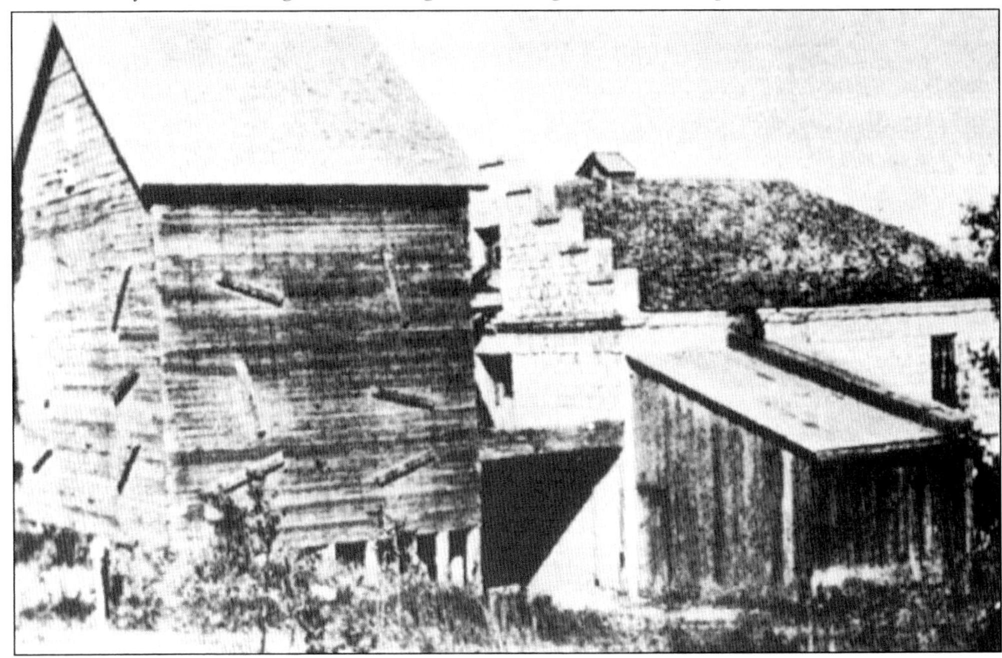
Stinnett Mill was built in 1874 by William H. Stinnett and John H. Orgain as a gristmill, grinding both corn and wheat. The original mill was a stone building two and half stories tall with a basement. It measured 30 by 40 feet with 22-inch-thick walls. In the early 1920s, a wooden granary was added to the front of the mill to dry and store grain. (Stinnett Mill Collection.)

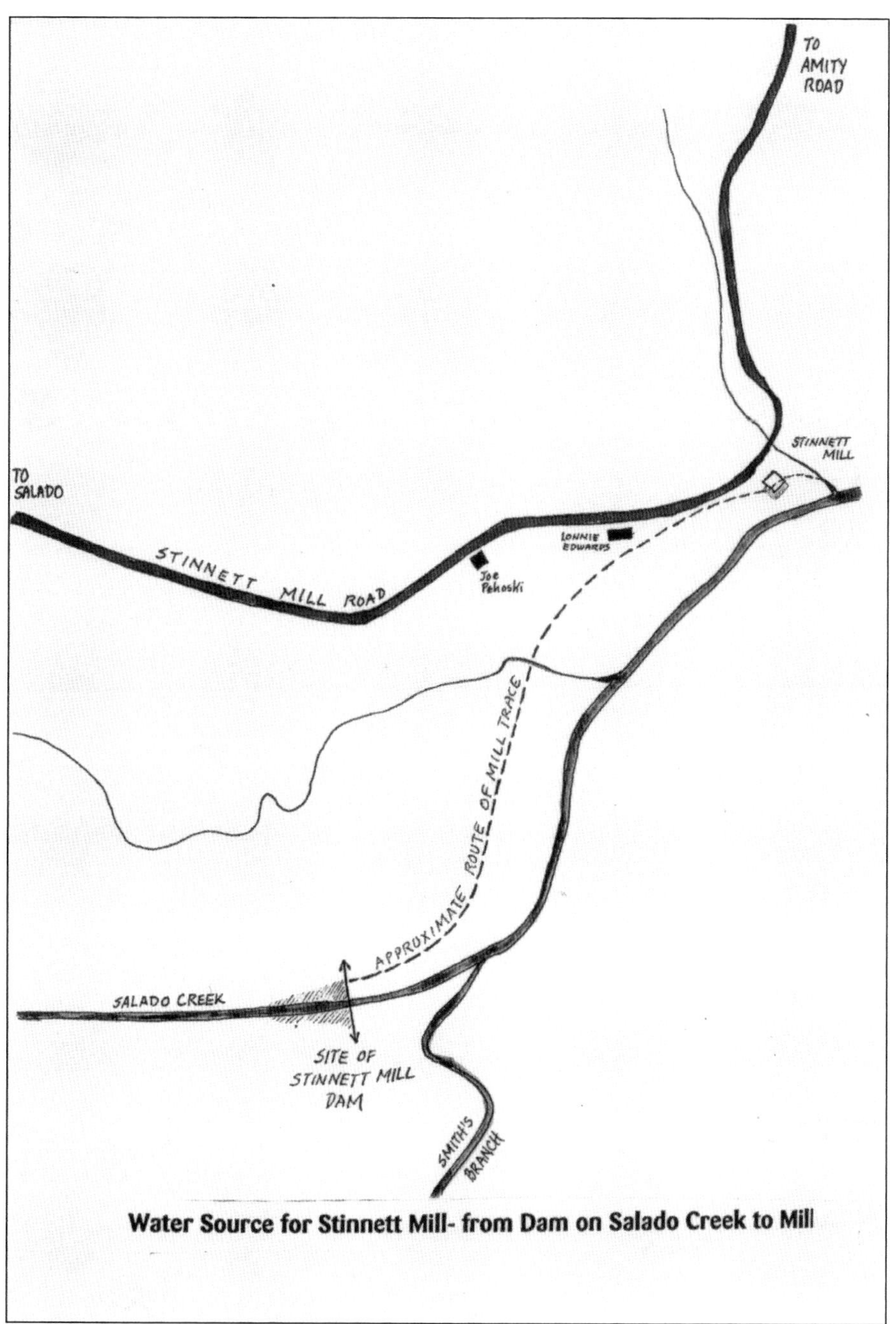

Water Source for Stinnett Mill- from Dam on Salado Creek to Mill

Of the eight mills along Salado Creek, Stinnett Mill is the only one floods have not destroyed. This is attributed to the fact that the mill is on higher ground about 900 yards from the dam that impounded water to supply the millrace. The above drawing clearly illustrates the mill's distance from Salado Creek and the approximate location of the millrace. Once the water entered the basement of the building through a stone archway, located in the creek-side corner of the mill, it poured into a shaft. It turned the turbine sitting in the bottom of the shaft and then was pumped or flowed out of the sluice at the opposite side of the building, where it returned to Salado Creek. (Dr. David J. Mikulencak.)

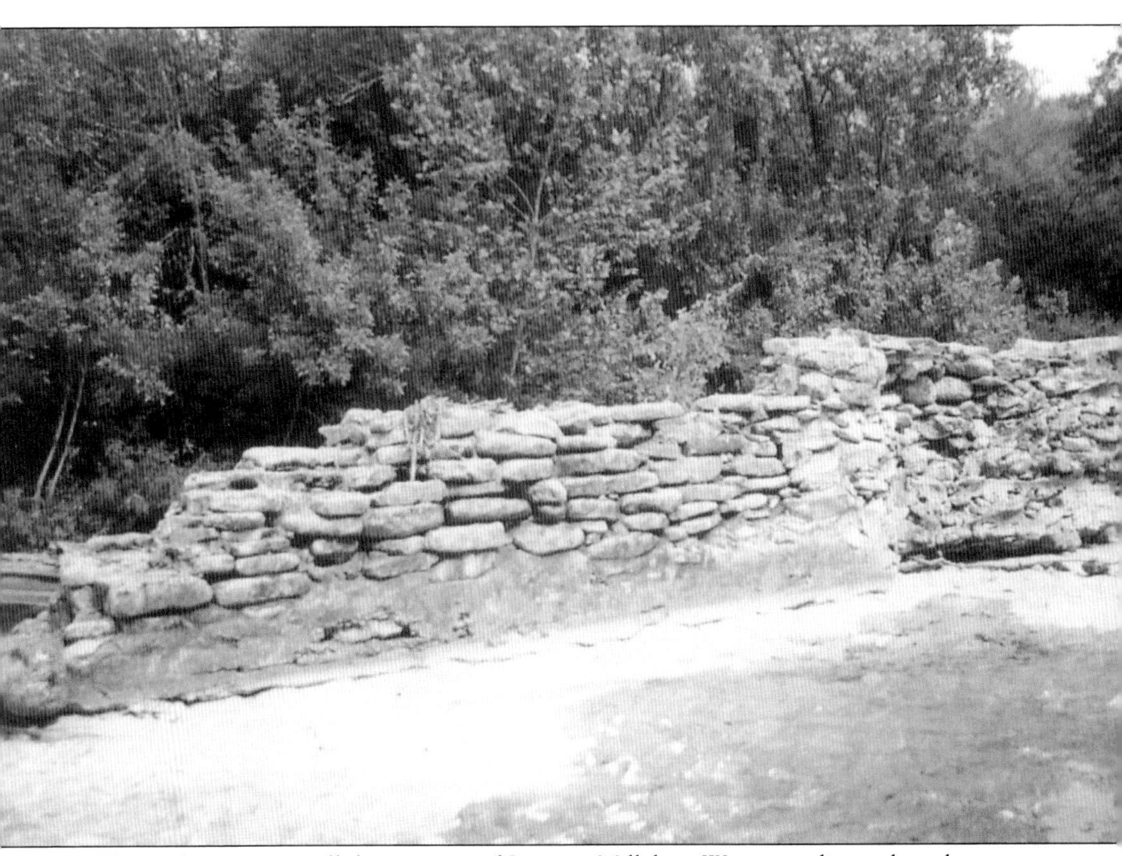

This rock structure is all that remains of Stinnett Mill dam. Water was diverted to a large masonry collection pool before it began its journey to the mill. This rock wall can still be seen from the Chisholm Trail low-water crossing in the Mill Creek subdivision. The collection pool has been silted in over the years due to flooding and erosion. (Dr. David J. Mikulencak.)

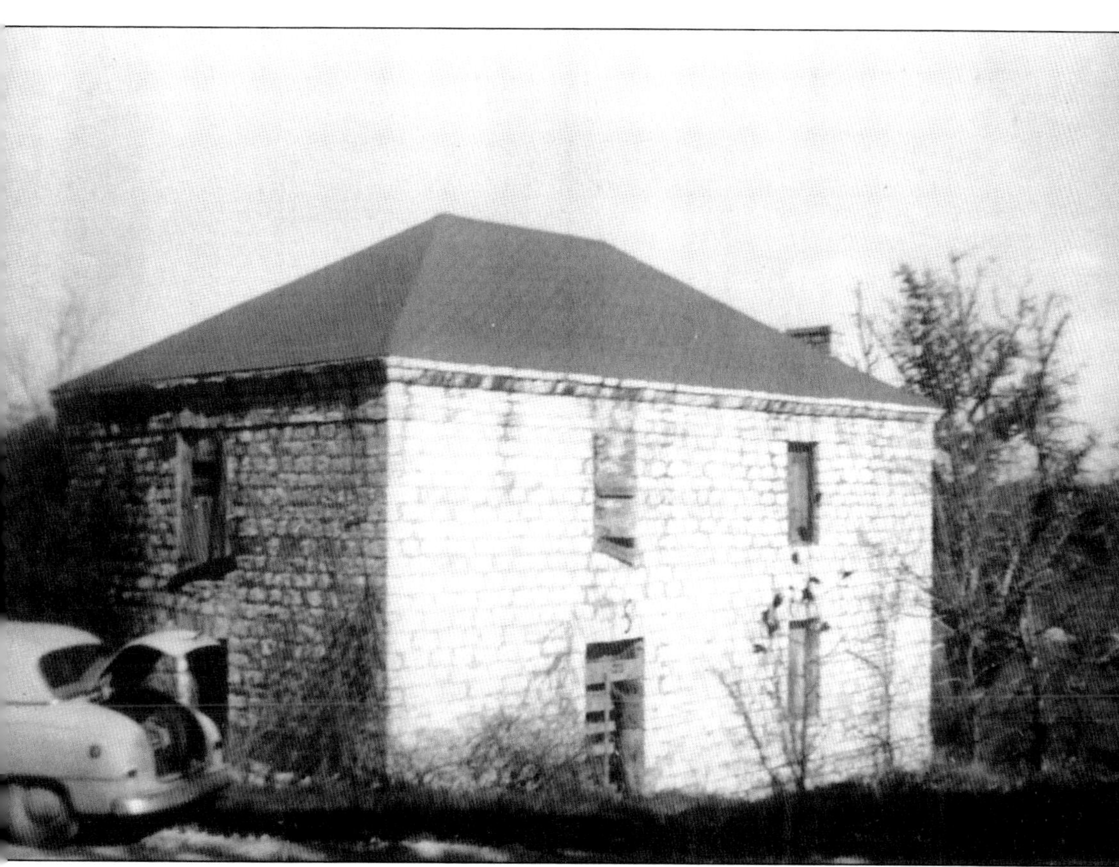

Stinnett Mill was in operation for over 50 years before suspending operation in the early 1930s. After sitting abandoned in an eerie silence for several years, the mill was purchased in 1945 by Ruth Berry Brown. Her plan was to convert it into a weekend home and preserve it as a monument to the history of the region. When Brown purchased the mill, its internal workings were still in place. One was a pit about 4 feet across and 30 feet deep. At the bottom was the turbine waterwheel that furnished power to the mill. At the south end of the ground-level floor sat a large fireplace, built of native limestone. Around 1968, artist Forrest Gist and wife Linda acquired the mill from Ruth Berry Brown. Gist planned to use the historic mill as his home and art gallery. The original rock facade was removed, and a new roof was added to begin conversion of the mill into the Gist residence. (Stinnett Mill Collection.)

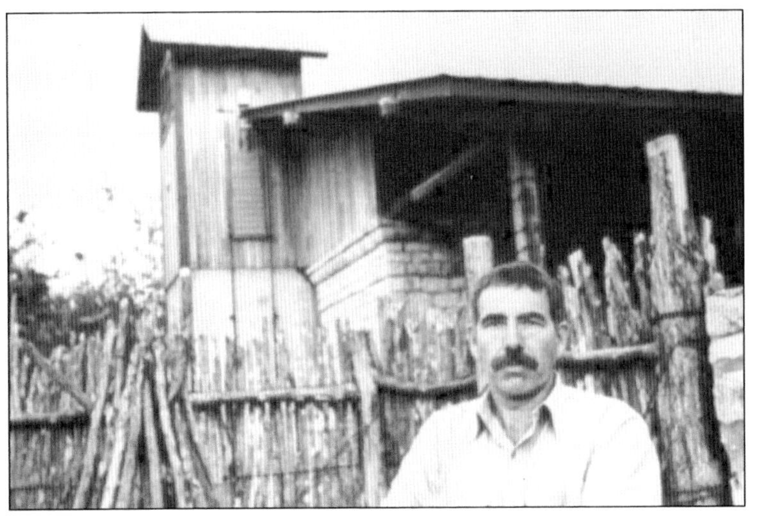

Forrest Gist poses for a photograph in front of Stinnett Mill after the restoration was complete. In addition to removing the rock facade, Gist added a functioning hand-operated pulley elevator to the front of the building. The elevator was adopted from the Johnson's Piano building in Temple when it was torn down. (Stinnett Mill Collection.)

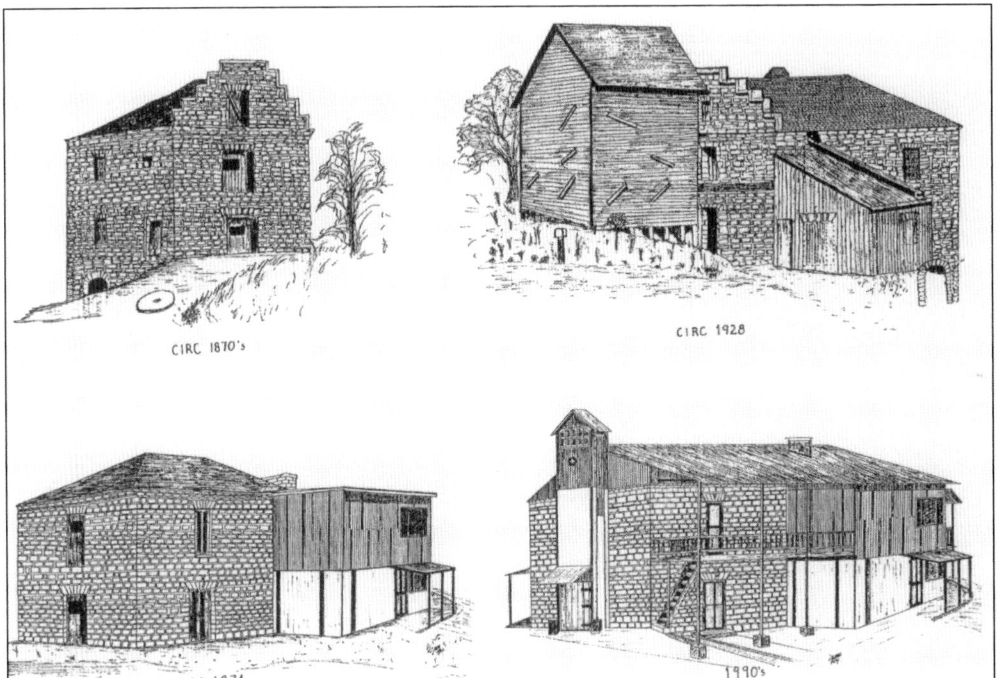

CIRC 1870's

CIRC 1928

CIRC 1974

1990's

These sketches show the evolution of Stinnett Mill over time. The 1870s sketch shows the original mill. The stone archway where the water that powered the turbine exited the basement is clearly seen in the lower left corner. The 1928 sketch shows the mill during its working years. The 1974 sketch shows additional living space, and the 1990s sketch shows the way the mill looks today. (Dr. David J. Mikulencak, artist and former owner.)

Stinnett Mill began operations over 150 years ago. The millrace and the waterwheel turbine are long gone. Other features of the mill have been modified to fit the needs of the various owners. However, the item that is still intact is one of the grindstones used at the mill. This grindstone is now imbedded in the stonewall fence in front of the mill. (Author's collection.)

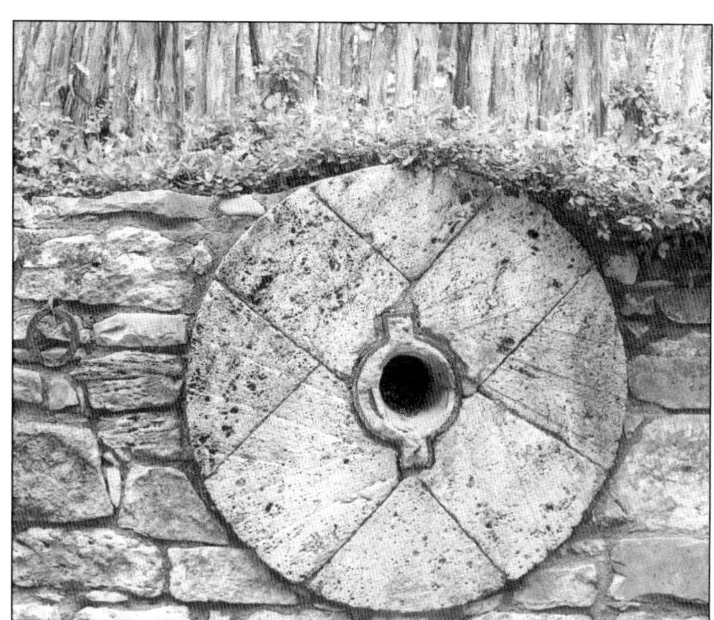

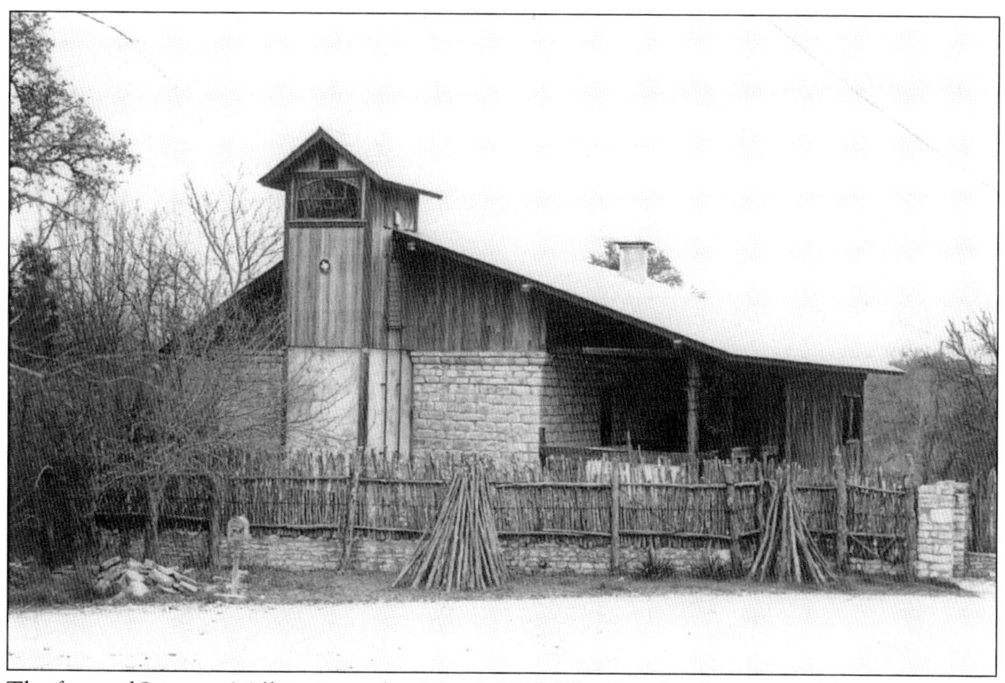

The front of Stinnett Mill is pictured in March 2015. The Gist family was careful to preserve the rustic appearance of the mill. When a second couple, David and Pat Mikulencak, purchased the mill, they added energy-efficient features and a more contemporary interior design. The David and Whitney Hill family now call the mill home. All families have been careful to preserve the history of the mill. (Author's collection.)

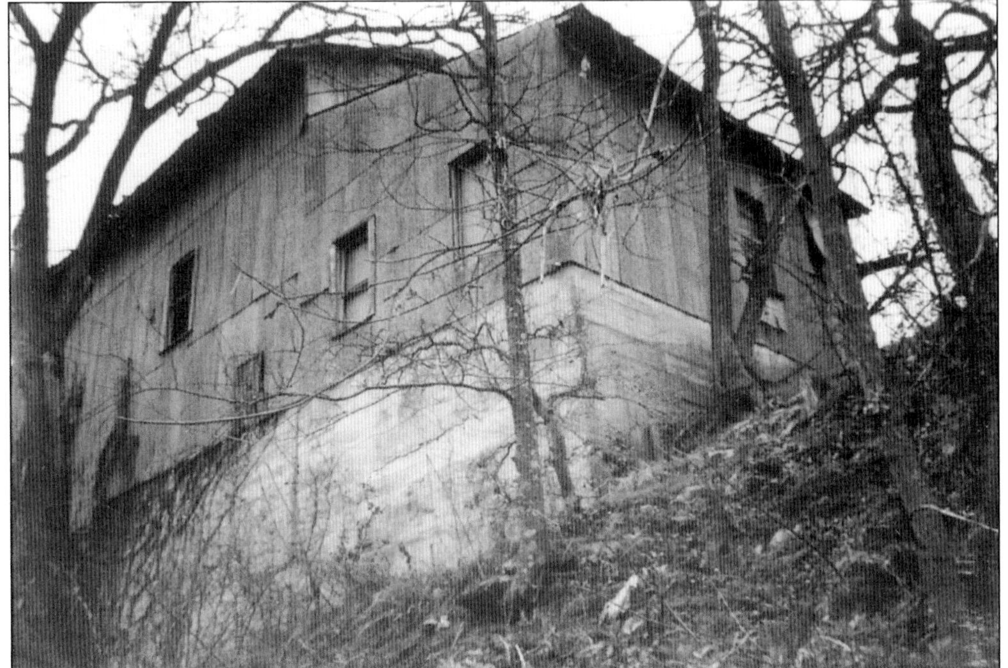

This 1995 photograph shows the remains of the mill constructed in 1880 by Isaac (Ike) V. Jones. The mill, located near Amity Road East, was built as a gristmill. When cotton became a dominant crop, Jones converted the mill to a cotton gin. In 1917, with the introduction of more modern ginning methods, the old converted gristmill gins could no longer meet the demands and were forced to close. (Stinnett Mill Collection.)

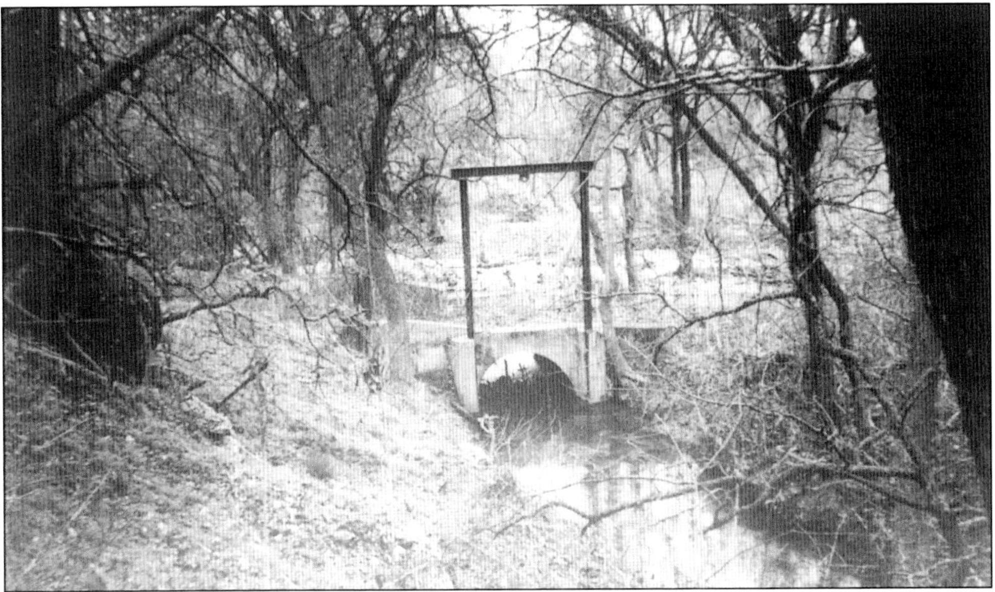

This is the sluice gate that started the water on its way to the mill. After the mill ceased operations, the half-mile-long millrace was used for swimming, and the millhouse became a bathhouse. The beautiful scenery around the mill site made the Ike Jones Mill location a popular recreational spot. (Stinnett Mill Collection.)

This is a view approaching the Ike Jones Mill, later cotton gin, from the south. The water entrance, shown in the lower right, is blocked with concrete cinder blocks. After the 1921 flood, the creek bank was littered with twisted machinery, including the mill hopper and the mill wheel, with gears and pulleys still attached to a drive belt to power the machinery. (Stinnett Mill Collection.)

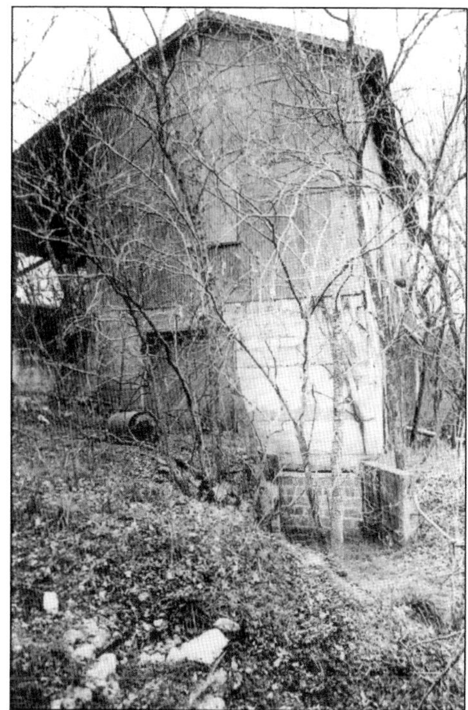

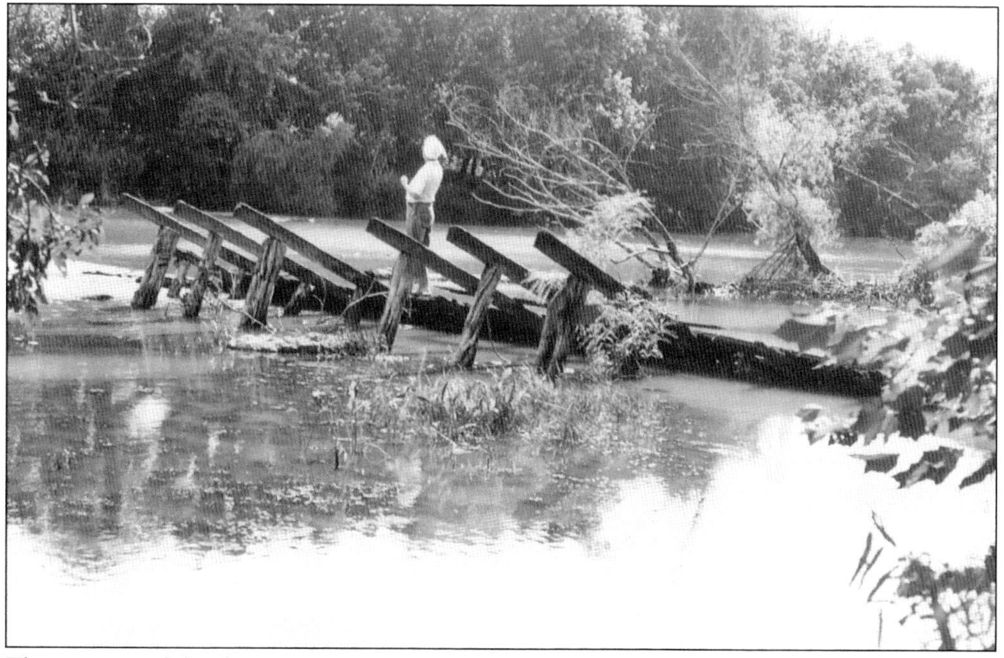

The remains of the Ike Jones Mill dam are pictured after the 1921 flood. This well-engineered dam was built with cedar posts sunk with an upstream incline and was located more than half a mile upstream from the mill. This location allowed a sufficient drop in elevation to make possible the free flow of water through the sluice. (Photograph by Alvin Thomas Jackson, Alvin Thomas Jackson Papers, di_10681, the Dolph Briscoe Center for American History, the University of Texas at Austin.)